ANIMAL LIFE IN FRESH WATER

ANIMAL LIFE IN IRISH WATER.

Animal Life
in Fresh Water

A GUIDE TO FRESH-WATER INVERTEBRATES

by

Helen Mellanby, M.A., Ph.D.

Formerly Demonstrator in the Department of Zoology
University of Sheffield

with a Foreword by

L. E. S. Eastham, M.A., M.Sc.

Professor of Zoology University of Sheffield

METHUEN & CO LTD

36 ESSEX STREET · STRAND · LONDON WC2

First Published September 29th 1938
Second Edition August 1942
Third Edition, Revised March 1949
Fourth Edition January 1951
Fifth Edition, Revised October 1953
Reprinted 1956

5.2
CATALOGUE NO. 3019/U
PRINTED IN GREAT BRITAIN
AT THE UNIVERSITY PRESS, ABERDEEN
AND BOUND BY HUNTER & FOULIS LTD, EDINBURGH

To K. M.

FOREWORD

By Professor L. E. S. EASTHAM
Professor of Zoology in the University of Sheffield

MOST books are written with the intention of supplying some particular need, but few end with such single purpose. Mrs. Mellanby's is no exception, for while the author planned this work to serve as a guide to the school pupil, which function it fulfils in an admirable way, it will also prove of value to the teacher, the university student and the amateur naturalist.

While it may be argued that it is not the function of the Universities to teach Natural History in the commonly accepted sense, it will always be the aim of Zoologists to know more about animals, what they are and do, where they live and why they live in particular environments. It is unfortunate, in view of the fact that the majority of students of Zoology enter the teaching profession, that the increasing load of instruction in morphology, physiology, cytology, genetics, evolution and the like frequently makes a personal study of animal life in relation to environment almost impossible. The fortunate ones visit the sea for a fortnight's course in Marine Ecology ; the others take posts in schools without even this respite and set about converting their academic learning to a school curriculum.

vii

The result is an undesirable and often slavish imitation of university method in the school classroom.

To know living animals is a first step to the ecological approach and to the analysis of environment which should occupy an important part in that branch of Zoology which has come to be known as "experimental." The university graduate whose work has perforce been restricted to the laboratory will find in this book a simple guide to the identification of the common animals of fresh waters with notes on their habits and distribution. To the schools it will serve equally well, for it is written in the most elementary terms and assumes on the part of the reader the minimum knowledge of animal structure and classification. The author brings together in easily readable form the results of an extensive survey of all groups inhabiting fresh waters. As a general rule only those characters which are visible to the naked eye or with the aid of a hand lens are used for purposes of identification. The pupil in searching for some of the more readily seen taxonomic features can thereby cultivate his powers of observation.

The simple line drawings with which the book is profusely illustrated are, in the majority of cases, original and specially drawn for the purpose from living specimens.

While fresh water is only one environment of many, its study has many advantages. No one in this country lives far from it, whether it takes the form of a mountain torrent, a horse pond or a water butt. This accessibility, coupled with the fact that fresh-water aquaria are easy to set up in a classroom, makes the study of fresh-water animals an attractive one to country and town schools alike.

PREFACE TO THE FIFTH EDITION

An opportunity has been taken of revising the nomenclature and in some instances the classification of the animals featured in this book. It is hoped that this will bring them more into line with modern usage. The occurrence of a number of species new to Britain also receives mention.

The references to more detailed works at the end of each section or chapter have been revised, and a number of recently published monographs, dealing especially with insect families and genera, have been added.

I should like to thank Dr. T. T. Macan of the Freshwater Biological Association's laboratory on Windermere, for his valuable assistance with this revision. I am also indebted to my daughter Jane for help with checking the necessary corrections in the text and index.

HELEN MELLANBY.

291, Great North Way
London, N.W.4.
September, 1952.

CONTENTS

INTRODUCTION

THIS book is intended to help all who are interested in fresh-water animal life to identify the animals which they find, and to discover more about their ways of living. I hope that it will be useful to people of school age as well as to more serious students of Natural History.

I have limited the scope of the book so that it deals only with Invertebrate animals, because there is a number of books on fresh-water Vertebrates (Fish, Frogs, Newts), whereas for information about invertebrates it is frequently necessary to consult original papers. Some of the common invertebrate groups of animals, such as Leeches and Flatworms, have previously been almost entirely neglected in books of this kind.

For the purpose of finding out to which group an animal belongs, a reasonably good hand lens is all that is required for animals larger than a millimetre ; animals smaller than this require a microscope ; most such animals will be found to belong to the Protozoa or to be young stages in the life-history of larger animals. With a very little experience it should be possible to place any animal in its correct group (phylum), after which reference should next be made to the appropriate chapter. Where there are only a few British species of any one class of animals, every species will be dealt with individually ; for example, the Leeches, of which there are eleven species. On the other hand, when any group (phylum) is large and its members difficult to identify, then only the various families or genera will be found ; i.e. the Arthropoda. The Protozoa, the Nematodes, and the Parasitic Flatworms are all groups which require specialists for their identification, so they are necessarily dealt with somewhat summarily here. The wholly microscopic groups (Rotifers,

I

Protozoa) are described in the two final chapters of the book because their study necessitates the use of a microscope and presents many difficulties.

The classification of the animals is mainly based on that found in " The Invertebrata " (by Borradaile, Eastham, Potts, and Saunders, 1932). This forms a framework for the text of the book. I have kept to the common name for any animal where one exists, the scientific name following in brackets. The sizes of animals are all given in centimetres or millimetres, these being the usual units used in biological works, but if you are not used to thinking in these units you should refer to a ruler.

In the case of the larger (macroscopic) animals, the straight line beside each figure indicates the natural size of the animal. For microscopic forms, the scale is given by means of a line which is magnified to the same extent as is the animal ; the actual length represented is given in " microns " (abbreviated to μ, pronounced " mew " ; 1μ is a thousandth of a milli-metre). An animal 100μ ($1/10$th of a millimetre) long is just visible to the naked eye as a tiny speck. Except where otherwise stated, all drawings have been made from live animals specially for this book by the author. The rest have been adapted and re-drawn except for Fig. 152, which was kindly lent by Dr. J. Smart.

In the preparation of this book earlier general works have been freely consulted. The following, which I found very helpful, will also be useful to the more advanced student, although they do not deal primarily with British forms : " Freshwater Biology," by H. B. Ward and G. C. Whipple (an American work) ; " Die Susswasserfauna Deutschlands," edited by A. Brauer, and " Ferskvandsfaunaen," by C. Wesenberg-Lund. Other less-advanced works have given much valuable information. These include " Aquatic Insects," by L. C. Miall ; " L'Aquarium de Chambre," by F. Brocher ; " An Introduction to Zoology," by R. Lulham ; " Biologie der Tiere Deutschlands," edited by P. Schulze ; " Das Susswasser-Aquarium," by E. Bade, and " Die Kleintierwelt unserer Seen, Teiche und Bache," by J. Hauer. It is impracticable to give a comprehensive list of more technical literature consulted on the various groups here, but a list of references giving detailed information on particular groups is printed at the end of most sections.

It is impossible to acknowledge the valuable help received from all of a large number of other Zoologists, but I must particularly mention the following : I am indebted to Professor L. Eastham for continual help and encouragement, and also for reading and criticising the very long insect section ; to Dr. Sidnie Harding and Mr. C. Oldham for reading the chapters and giving their help with the Crustacea and Mollusca respectively ; and lastly to Kenneth Mellanby for giving much time during the whole preparation of the book to the arranging of the chapters, collecting of live material, and helping with the figures. But for his assistance I am sure the book would never have been finished.

CHAPTER 1

LIFE IN FRESH WATER

THE animals which live in fresh water belong to many different groups of the animal kingdom. We are not concerned in this book with those animals which have backbones (Vertebrates), but it will be found from observation that the bodies of the others (Invertebrates) are built up on a number of very different plans. For example, there is very little resemblance between the structure of, say, a water-snail and that of a water-beetle. Animals are classified into large groups (phyla), members of each group having certain fundamental structures in common. Those placed in one big group may have many features different from those in another group. It is unusual to find that all the members of one group live in fresh water ; it is much more common to find that a few species only inhabit inland waters, while the rest live in the sea or on dry land. In some cases only a part of their lives is spent in fresh water ; this is a common state of affairs among aquatic insects.

All invertebrates are cold-blooded ; that is to say, they do not regulate the temperature of their bodies like mammals or birds. This means that their temperatures are approximately that of their surroundings, whether they are in air or water. The surrounding temperature fluctuates between night and day and winter and summer in this country, so the body temperature of the animals goes up and down. The variation in temperature is less in water than on land, the amount of variation decreasing with the increased size of the water expanse, since large volumes of water take a long time to heat up or cool down.

In this country the number of animals in ponds and streams

is greatest during the summer months. There are two reasons for this ; first, with increase in warmth the animals become more active, grow more rapidly, and breed more quickly. Many of them are quick breeders, so that a few weeks of warm weather greatly increases their numbers. The second reason is that a large number of the animals spend the winter months in a resting, cold-resisting stage, among the mud at the bottom of ponds and lakes ; they are then very difficult to find.

Water is a much more buoyant medium in which to live than is air. Water exerts an upthrust on the bodies of animals so that they do not have to support all their own weight. Soft-bodied creatures like Hydra (p. 15) would be unable to expand themselves in air, but can do so with ease in water. Small animals are able to glide through the water by means of hundreds of little lashing tails (cilia), such as are found on the surface of Flatworms or many Protozoa. Such a method of progression would be difficult on land except where films of water existed. Many animals make use of the surface film of water to support all their weight. For instance, pond skaters (p. 148), with their very long legs, skim along on the surface, while others, like mosquito larvæ, snails, and flatworms, hang or glide upside-down in the water, attached to the surface film which supports their weight.

Fresh water always contains some dissolved oxygen ; it is this dissolved gas which most of the animals breathe. Cold water which is well stirred up by passing over waterfalls contains much oxygen. Warm, stagnant water contains less. Different animals have different oxygen requirements, which often helps to explain why any particular species is only found in one definite type of fresh water. Animals of very small size do not need special organs for obtaining oxygen ; the gas finds its way (by diffusion) through the surface of their bodies from the surrounding water. Larger animals, like some insects and Crustacea, whose surface area is inadequate for respiration, would not obtain an adequate supply by this method, so they may have special gills, or they may swim to the surface to breathe the oxygen of the air. On the other hand, animals which normally live in badly oxygenated water, such as the mud at the bottom of a pond, may have a special red substance in their blood (hæmoglobin) which combines with oxygen from the water. This red pigment is found in such unrelated species as the larval (young) stage of a midge (*Chironomous* larvæ or " blood-worms "), a water-snail (*Planorbis*), and a small segmented worm (*Tubifex*). In sunlight water-plants give off considerable quantities of oxygen, and this helps to increase the amount present in ponds during the daytime.

All animals contain in their bodies a large proportion of water which is necessary for life. Land animals have to conserve this water unless they live in very damp situations. This is done by the skin being to a large extent waterproof. Fresh-water animals do not need this protection, which makes many of them very susceptible to drying when taken out of the water. Ponds, however, may often dry up gradually during hot summers. When this happens the animals retreat into the mud, and many go into resistant resting stages, or lay drought-resisting eggs. Such eggs and resting stages can be distributed to nearby ponds by the wind, or to more distant ones on the feet of birds. When favourable conditions return the pond will become re-populated from the development of the remaining resistant stages.

The food of fresh-water animals consists of either plants or other animals, or a mixture of both. Some species, like the large beetles (p. 156), water-bugs (p. 142), and young dragon-flies (p. 124), are notorious carnivores. It is well to learn their habits before placing one of them in the same container as another valuable specimen ! Some animals will eat up the dead bodies of others, and are often referred to as scavengers. Some feed by sucking the blood of their victims, or by living inside them, without necessarily doing any great bodily harm ; they are known as *parasites*, and their victim as the *host*.

The egg-laying habits of fresh-water animals are various, depending on the type of water inhabited. In streams and rivers the eggs must not be carried down to the sea, so they are laid among damp earth by the edge (some worms), in holes in plants (*Dytiscus* beetle), in jelly masses or cocoons attached to plants or stones (most snails), or they are carried about by the parent (water-fleas). The young hatch from the eggs in a suitable stage in development to be able to hold their own against the current, or they may continue to be carried about by the parent until they reach such a stage. In ponds special resistant eggs may be laid to preserve the species during bad conditions (e.g. Crustacea), in addition to ordinary eggs.

Notes on collecting

If you collect animals from streams, ponds, and ditches, you will soon observe that some of the animals live in almost any kind of fresh water, whereas others are only to be found either in ponds or streams, but not both, unless a stream is so slow-running as to be almost a pond. In looking for specimens the undersides of solid objects, such as stones or floating pieces of wood, should always be examined. This is particularly the case in swift-flowing streams or rivers where most of the living things will shelter from the current. In many instances they will be found clinging to the under-surfaces of stones or to the stems and leaves of submerged plants.

Collecting apparatus need not be elaborate for collecting material from ponds and rivers. Screw-topped jars and small glass tubes with corks are useful for carrying home living specimens. Many animals, more particularly larger species of insects and Crustacea, travel better amongst damp weeds than they do in water. For the actual collecting a small net of fine strong mesh, with a brass framework to which can be attached a stick, and a penknife for dislodging creatures from stones, are all that is essential. It is a good plan to carry a white enamel pie-dish into which you can place the contents of the net or the stones which you take from the middle of a stream ; then, as often happens, the animals on the stone may swim off, when they will be easily seen against the white back-ground ; a pipette or fountain pen filler is then the best instrument with which to remove small animals to another container. If a handful of water-weed from a pond or stream is placed in the pie-dish with some water for a few minutes, all kinds of animals will swim out of the weed and are then easily transferred to a tube. Quite a large number of animals live in mud or sand under water. Larger forms may be obtained by scooping up the material on to the bank and then watching it for any signs of movement. Insect larvæ and worms are easily found this way, while molluscs (fresh-water mussels, etc.) will be recognised by their shells. When you are examining a pond it is important to remember whether you found the animals on the surface, swimming in the water, or in the mud at the bottom. When you collect animals from a stream you will find that you get different types living on the stones which are in a strong current of water from those which live under stones in comparatively still backwaters. Again, a totally different community of animals will be found in the submerged mud at the edge of a stream. A pebbly stream with a strong current contains few animals because the stones are too small to afford protection. When you wish to obtain microscopic animals you have to take samples of water and mud from likely looking places, then examine them under the microscope at home. It is difficult to get animals from the large lakes without special apparatus, though the stones and vegetation at the edge may easily be examined. If you happen to go out in a small motor-boat or a yacht it may be possible to drag a tow-net through the surface water and so obtain floating forms.

When you get the material home, any containers, such as jam jars or pie-dishes will do for keeping the animals in for weeks or even months, but they must not be allowed to dry up. You will get much more interesting results from keeping a number of small separate aquaria of this type and examining them two or three times a week, than you will get from one beautiful large glass tank, though the latter is useful for keeping a stock of snails, water-fleas, and other things which do not eat each other. Animals should not be overcrowded if they are kept in small dishes. It is very important that there should always be a large surface of water in contact with the air in proportion to the volume of water, as otherwise the animals may die from oxygen lack. It is usually wise to cover up small dishes with a piece of flat glass or any other suitable object because if the water dries up you will lose your specimens. Excessive light and too high temperatures are harmful to many animals and should be avoided. Small specimens may be pipetted on to a slide for examination under the ordinary microscope. Jars and tubes containing samples of mud, pond water, or aquatic plants should always be examined every few days. The more conspicuous members of the original fauna may die out, but probably will be replaced by a great increase in numbers of all sorts of other less-conspicuous creatures which were originally present in rather small numbers ; in the end they may die too and there will only be Protozoa and Bacteria left.

For further general information on Life in Freshwater see
1. Life in Lakes and Rivers, by T. T. Macan and E. B. Worthington. London, 1951.
2. Biologie der Süsswassertiere, by C. Wesenberg-Lund. Wien (1939).

CHAPTER 2

FRESH-WATER SPONGES

NEARLY all sponges are sea animals, but one family, the Spongillidæ, live in fresh water. There are five British species, two common ones being the River Sponge (*Ephydatia* (or *Spongilla*) *fluviatilis*) and the Pond Sponge (*Euspongilla* (or *Spongilla*) *lacustris*). Both these species form greenish or yellowish encrusting growths over the surfaces of stones and water-plants. The colour depends on whether the sponge is growing in a good or poor light, because the greenness is due to tiny one-celled plants similar to those which live inside Hydras (p. 15). These plants only develop their chlorophyll (the green colouring matter in plant cells) when they are exposed to light. The River Sponge is common in many places, it forms flat encrustations on the underside of stones, when its colour is yellowish or dirty white ; green specimens are found on plant stems or pieces of timber which are exposed to the light. Rivers and canals are its chief habitats. The Pond Sponge is found in still water only. It has a similar encrusting habit, and is characterised by a large number of finger-shaped projections on its surface (Fig. 1). It is generally a brighter green than the river sponge and it has a coarser texture.

These sponges live well in captivity, but they are rather dull objects to study unless their microscopic appearance can be examined. The surface of the encrusting growth is covered by very tiny holes through which a current of water passes inwards. There is a number of larger holes or *oscula* which are usually easy to see, and through them a current of water passes out. You can see this current if you place a piece of plant, encrusted with a sponge growth, in a dish of water under a dissecting microscope and add to the water a small amount of carmine particles. When the sponge has recovered from being transferred from the aquarium it will start to produce this current of water through its body and the carmine particles will be drawn into the small holes and pushed away from the large ones. The current is made by cells which line a number of communicating compartments inside the body. These cells each possess a long tail process (flagellum) which waves about in such a way as to produce a current of water which is directed outwards through the large openings ; water is sucked in over the surface to replace this. The current of water is essential to the sponge because it brings with it small food particles which are taken up by the flagellated cells, a good supply of oxygen to the tissues, and a supply of the substance silica out of which the sponges build a skeleton. If a small piece of sponge is squashed and examined

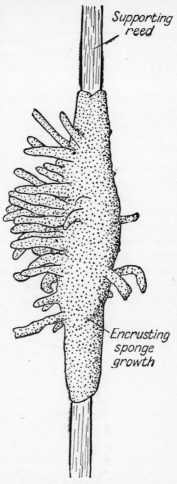

FIG. 1.—Pond Sponge (*Euspongilla lacustris*). After Bade.

under a microscope the most obvious thing will be a large number of transparent pointed rod-like structures which are called *spicules* (Fig. 2). These are made of silica

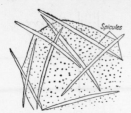

FIG. 2.—The River Sponge (*Ephydatia fluviatilis*), part of skeleton much magnified.

extracted from the surrounding water; their presence in the sponge body supports it to some extent, and gives the sponge a crisp feeling. The spicules form the skeleton, but they have no definite arrangement. If you are in any doubt as to whether a certain specimen is a sponge or not you should examine it for spicules. (The ordinary bath sponge belongs to a different family which has no spicules but only a tough skeleton made of silky material with which we wash ourselves.)

The amount of water which goes through a sponge is considerable, and for this reason it is necessary to keep them in a large aquarium and to have each specimen (if it is large) in a separate container, otherwise they do not obtain enough food or silica to grow. Sponges can regenerate very easily, and they are not killed if part of the growth is cut off. They do not move about, but remain fixed to some surface all their lives, the area which they cover enlarging with growth.

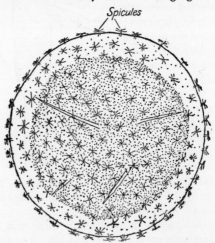

FIG. 3A.—Gemmule of River Sponge (*Ephydatia (Spongilla) fluviatilis*).

Towards the end of the summer the body contains a large number of dark brown bodies about the size of a pin-head. These each contain a number of sponge cells surrounded by a horny protective covering and a layer of special short spicules (Fig. 3A). The main sponge growth dies down in winter, being replaced the following spring by a growth which starts from the sponge cells enclosed in the brown bodies or *gemmules*. These gemmules may be made to germinate in the laboratory by placing one or two in a small dish of water. After a few days they produce a small sponge.

Sponges also have a sexual method of reproducing themselves. The egg and sperm cells develop inside the sponge and the ripe sperm are set free into the water ; they are drawn into another sponge by the water current where

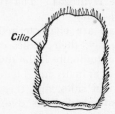

Cilia

FIG. 3B.—Larva of Fresh-water Sponge (*Spongilla*), natural size 0·3 mm. long. After Wesenberg-Lund.

they fertilise the eggs. Development of the egg takes place inside the body of the parent until it has reached the stage of being a *larva*. The larvæ are ciliated all over, they escape through the oscula and have a brief free swimming existence of about twenty-four hours. They then attach themselves by one end to some object (Fig. 3B) and undergo a very curious metamorphosis into a tiny sponge. During the metamorphosis the outside ciliated cells migrate inwards, coming to lie in the inside of the internal cavities, where they become the flagellated cells responsible for producing the current of water. Larvæ are produced in summer.

Almost any fresh-water sponge which you examine will be found to harbour a number of other animals, some of them parasites like the *Spongilla Fly* (p. 152), others just using the sponge body as a good place to shelter.

For further reference see
Stephens, J. (1920), The Freshwater Sponges of Ireland, *Proc. Roy. Irish Acad.*, **35**, 205.

CHAPTER 3

HYDROIDS. THE FRESH-WATER CŒLENTERATES

THIS phylum of animals is represented by many forms in the sea, such as the Jelly-fishes, Corals, and Sea-firs, but there are only a few fresh-water species all belonging to the same subdivision called the Hydrozoa. The two principal genera are *Hydra* and *Cordylophora*.

Cœlenterate animals are characterised by having radial symmetry, that is to say, their bodies can be divided into two similar halves longitudinally if the division is made along any diameter. (Most animals are bilaterally symmetrical; they have left and right sides, and there is only one plane along which longitudinal division divides them into two similar halves.) Radial symmetry is usually considered to be connected with a sedentary life, thus most cœlenterates are fixed or grow in fixed colonies for at least part of their lives. The body of an individual animal is like a hollow sac with a mouth opening at one end surrounded by thin finger-shaped processes, the *tentacles*. The mouth opening is often slightly raised on a prominence between the tentacles; food is taken in through this opening, and waste material is expelled through it as there is no anus. The body wall of the animal is made of two layers of cells only which are differentiated to produce all the types of cells found in the animal. The inner layer lines the large space in the middle of the body which serves as a gut cavity, these cells have digestive and absorptive functions. There is no body cavity distinct from a gut cavity, as there is in all other animal phyla, except the Protozoa, Sponges, and Flatworms. The surface of the tentacles is studded with groups of peculiar stinging cells known as *Cnidoblasts*. These are used as organs of defence and for catching prey. The active principle of these cells is a sticky thread sometimes barbed at its base which is shot out as a result of pressure on the walls of the structure which contained it. An irritating fluid may be ejected at the same time, the result is that live prey may be paralysed by the irritating fluid entering a puncture in their skin, and their bodies entangled in sticky threads, while the tentacles bend over and push the food into the mouth.

Genus Hydra.—A detailed description of the anatomy of Hydra will be found in any zoology text-book, so that only a very brief account will be given here. The body is a simple sac with a mouth and tentacles at one end (Fig. 4A). There are probably three British species, one green (*Hydra viridissima*), one usually yellowish-brown but sometimes greyish (*Hydra vulgaris*), and a third generally grey but sometimes reddish-brown (*Hydra oligactis*). The colour is in the inner layer of cells only, the outer layer is clear white, as can be seen very clearly by examining a contracted specimen under a low-powered microscope. Minute one-celled plants (*zoochlorellæ*) in the cells of the inner layer are responsible for the colour, these being either green, brownish, or grey, each species of Hydra having its own species of plant. *Zoochlorellæ* are included in the developing eggs, and are thus passed on from one generation to another. The plants are not parasites but live in symbiosis with the Hydra. They gain a suitable situation in which to grow and get a supply of nitrogen from the Hydra's food; on the other hand the Hydra obtains oxygen from the by-product of the plant's photo-synthetic activities. Hydras can contract into a small rounded knob or expand to very many times their contracted length. This is one of the reasons why they are difficult to find in nature, because unless you actually observe them hanging extended from plants in the water, they contract up so much when you remove the plants to examine them that they are not easy to find. The best places to look for them are ponds, ditches, and fens where there is a carpet of duck-weed on the surface, or many floating water-lily leaves. Careful examination of the under-sides of the water-lily leaves will probably result in your finding a number of specimens. In the Norfolk Broads about

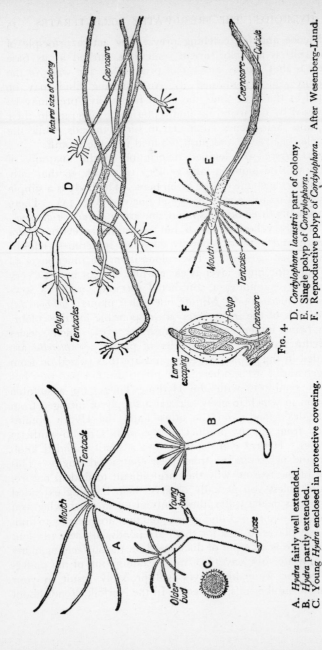

Fig. 4.

A. *Hydra* fairly well extended.
B. *Hydra* partly extended.
C. Young *Hydra* enclosed in protective covering.

D. *Cordylophora lacustris* part of colony.
E. Single polyp of *Cordylophora*.
F. Reproductive polyp of *Cordylophora*.   After Wesenberg-Lund.

July and August, one leaf may have four or five specimens attached to it, they are so common. Another method is to take home a quantity of duck-weed and place it in a jar of water. On examining the jar next day the Hydras will very probably be found attached to the glass sides or hanging down from the roots of the plants.

Hydras are normally solitary animals living temporarily attached to plants, or some support, by the end opposite the mouth. They are able to move slowly by turning somersaults along some supporting object, they are then alternately attached by the base and the tentacles. If they are able to get a lot of food and the temperature conditions are favourable, they reproduce themselves (asexually) by forming buds. The body wall at one point on the side of the parent becomes pushed out to form a small knob, at the end of which a mouth and tentacles develop. The mouth leads into the main cavity of the parent Hydra for some time, but eventually a constriction occurs at the base of the new Hydra and it separates from the parent. Under particularly good conditions one Hydra may have five or ten buds so that the whole forms a temporary colony. Food consists of small worms and small Crustacea, such as water-fleas, which are captured by means of the stinging cells on the tentacles. Hydra will only thrive and multiply in an aquarium if it is provided with plenty of food, and when small they must be provided with small water-fleas, otherwise they will be unable to eat them.

As well as reproducing by budding Hydras reproduce sexually. An individual is generally hermaphrodite and produces both male and female cells. The male cells, the sperms, are developed in small bulgy thickenings on the surface of the animal usually fairly near the mouth end. When the sperm cells are complete they are liberated into the surrounding water. The eggs are produced in similar thickenings which generally develop nearer the attached end of the animal. Only one egg is produced in each thickening (which is really an ovary), and when this is ripe it is fertilised while still in the body of the parent by sperm from the surrounding water. After fertilisation it develops into a two-layered ball of cells, the outer of which secretes a thick

2

protective covering. At this stage (Fig. 4c) it is set free into the water, and it is resistant to unfavourable conditions such as cold or drought. Normally it rests for several weeks before proceeding with its development, during which time it may be carried some distance from the parent, or occasionally a long distance if it should happen to be picked up in the mud on a bird's foot. Under suitable conditions development begins again inside the protective covering, the latter is ruptured, and a small Hydra emerges which has developed a mouth and tentacles. This feeds and grows until it is large enough to develop buds and reproduce sexually. Though Hydras are normally found in still or very gently flowing water, they may sometimes be found under stones in streams where there is a considerable current.

Hydras frequently have a small barrel-shaped parasite (*Trichodina*) which runs up and down the surface of their bodies. Sometimes these are present in great numbers ; they belong to the Protozoa and are mentioned on page 285.

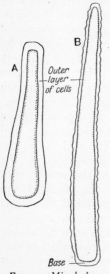

FIG. 5.—*Microhydra.*
(A) contracted.
(B) extended.

Genus Microhydra.—One species of this genus is found on the surface of stones in running water. The animal is like an ordinary Hydra without tentacles and it creeps slowly along or is fixed temporarily by its base (Fig. 5). Buds may be formed giving rise to a small colony, and apparently some of these produce another kind of individual like a small jelly-fish which separate off from the parent and swim freely ; they are of opposite sexes, and the fertilised eggs produced by them develop into the hydra-like individuals. These creatures are very small, only 1-2 mm. long, and they do not seem to be common.

Genus Cordylophora.—This is a type of cœlentrate which forms definite colonies like those of the closely related sea-firs. A colony consists of a large number of individuals each very much like a Hydra in structure which are joined together by a system of branching tubes (Fig. 4 D, E). The colony is produced in the beginning by the budding of one individual, but the buds grow long stalks before they form new *polyps*, as the hydra-like individuals are called. The branching tube connecting the polyps is called the *cœnosarc*. The cells covering it produce a transparent, hard, brownish material which helps to support the colony. The polyps are pinkish-white. They have numerous tentacles irregularly arranged, some coming off near the mouth, others lower down. These have many groups of stinging cells as in Hydra, and the food caught by each individual is used for the good of the whole colony. At certain times of the year special buds arise which have only a short stalk (Fig. 4F). These produce eggs or sperms inside a closed sac instead of developing into the ordinary tentacled polyp. Later an opening is produced at the end of the sac. Presumably the sperm escape through these and enter a sac containing eggs which are then fertilised. The eggs develop into elongated ciliated larvæ which are set free into the surrounding water. They fix themselves to some suitable object and grow into a new colony. The only species of *Cordylophora* in Britain is *C. lacustris* which is common only in certain places such as the Norfolk Broads and several estuaries including the London docks. It grows best and the polyps are largest in slightly salt water, but it is also able to live in completely fresh water. The colonies form tangled masses on the submerged parts of bridges and piers ; if covered with mud they are difficult to see. It is common to find other animals sheltering among the colonies such as small worms, insect larvæ, and Crustacea like *Corophium* (see p. 109).

CHAPTER 4

FLATWORMS (Platyhelminthes)

THE Flatworms are a group of animals which seldom receive much attention in Natural History books. Some of the larger free-living forms are very common in ponds and streams, while parasitic kinds may be of practical importance in that they cause disease in other animals such as sheep. As a phylum they are more advanced in structure than are *Hydra* and its allies, but they are less highly organised than the true segmented worms. The bodies of flatworms are usually flattened ; they are unsegmented, but they possess bilateral symmetry (i.e. they have a definite head end, and a line drawn through the middle of the body from head to tail divides the animal into two similar halves). There is only one opening of the alimentary canal, which is the mouth ; through this food is taken in and undigested matter is passed out. Many parasitic forms living in the alimentary canal of other animals have no mouth or alimentary canal ; they absorb digested food from their host, over the surface of their bodies. The internal anatomy of all flatworms is simple except in the case of the reproductive organs. Nearly all have both male and female organs in the same animal (hermaphrodite), and every individual therefore lays eggs when it is mature, instead of this process being restricted to females. As a result, complicated structures may arise in the reproductive system which are there to ensure cross-fertilisation of the eggs, by preventing self-fertilisation.

The phylum of flatworms is divided into three groups : the *Turbellaria*, the *Trematoda*, and the *Cestoda*. The Turbellaria are free-living animals found in fresh water, the sea, and on land. Most of the flatworms described in this book are Turbellarians. The Trematodes are with few exceptions parasites, living on, or in, the bodies of other animals. The adults of some are found attached to the gills of fishes. Young (larval) stages of others swim about in ponds and rivers for a short time ; they may sometimes be seen if water is examined under a microscope. The Cestodes are all internal parasites with no free-living stage in their life except the egg. The adult Cestodes live in vertebrate animals, generally in the alimentary canal ; a number of different kinds occur in fish. Other phases of the life-history may also occur in fish ; for example, there is one immature stage of *Schistocephalus gasterostei* which is found commonly in the body cavity of sticklebacks where its presence is very obvious because it makes a large bulge on one side of the animal.

(a) Class Turbellaria.—Free-living ciliated Flatworms

The fresh-water Turbellarians are found in any type of water from cold mountain streams to small stagnant ponds, but each species may be found only in one particular kind of situation.

Triclads.—The larger species of Turbellarians mostly belong to the group known as Triclads. They have a much-branched alimentary canal divided into three parts (see Fig. 6B) which gives the group its name. The branches of the gut may be easily made out in light-coloured specimens ; in black forms it is hidden by the dark pigment. The size of Turbellarians varies from about 1 cm. to 4 cm. in length when they are extended. Their breadth is considerably less than their length. When at rest they are found on the undersides of stones, floating leaves, or curled up among submerged mosses. They look like irregularly shaped lumps of blackish, brownish, or white jelly. Brown specimens are easily confused with leeches when seen for the first time. If they are carefully removed with a knife and placed in a dish or tube containing water they will usually begin to move by passing a series of muscular contractions down the body, or by extending themselves and then gliding by ciliary action slowly round the container. They often glide upside down along the surface of the water, their weight being supported by the surface

film.   When they are moving they show characteristic shapes
which will enable you to know which species you have found
after some practice.   Turbellarians can be distinguished from
leeches by their having no suckers on the under-surface of the
body though they are able to cling to surfaces by pressing
their bodies against them, and by the lack of segmentation
(compare with leeches on p. 58).

There are only nine different kinds of Triclads which
occur in fresh water in Britain, but some kinds are so common
that it is possible to find as many as thirty-two individuals
under one stone (*Dendrocœlum lacteum*) or twenty individuals
on the underside of one water-lily leaf (*Polycelis nigra*).

Drawings of the various Triclads are on page 24 ; these
will help you to decide on the species which you have found.
Each species has certain features shared by all.   The body
is flattened dorso-ventrally ;   it is approximately parallel
sided when the animal glides, and the tail is usually pointed.
The colour varies from pure white through grey and brown
to black.   In pale specimens if the gut contains coloured
contents then the whole animal will appear to be that colour
because the gut branches all over the body.   Thus if an
individual has just eaten a green caddis pupa (see p. 192) the
flatworm may appear at first sight to be green all over.   The
mouth is on the underside in the last third of the body, it is
therefore nearer the hind end than it is to the head.   The
mouth leads into a muscular sac (the pharynx) which can be
protruded from the body, this in turn leads into the gut which
divides into three main branches.   Each branch has many
smaller side branches which all end blindly, so that the only
way of getting rid of undigested food is through the mouth.
Behind the mouth in the mid-line is another opening through
which the cocoons are laid.   The surface of the body is
covered all over with tiny protoplasmic hairs or cilia, these
are in constant motion and cause the animal to glide slowly
over surfaces.   They may be seen in living specimens under
the microscope.   Slime glands are present just beneath the
surface, and the slime which they manufacture is secreted
on the outside of the flatworm ;   this serves to lubricate the
surface over which the animal is gliding.   Some pigment is

usually visible in the skin, but the quantity varies in two individuals of the same species. The cells of the skin contain in addition to pigment curious rod-shaped bodies which are crystalline ; they are called Rhabdites. Their function appears to be unknown. The movement of the cilia creates currents in the surrounding water, and it is this characteristic which gave the name Turbellaria to the group.

Eyes are possessed by all the British Triclads, their number and arrangement being a good guide to the identification of a species. Some have tentacles at the head end which appear to serve as a kind of sense organ ; it has been suggested that they may help in searching for food.

All the Triclad Turbellarians are carnivorous, eating small animals of almost any kind but particularly insects and crustaceans. If the prey is much smaller than the Triclad it is taken whole into the mouth, if it is larger, it is first wrapped up in slime, then the muscular sac (pharynx) is protruded through the mouth and it sucks off pieces of the prey, which are then taken into the mouth. Small animals which have recently died may also be eaten. If a piece of meat attached to string is left in a suitable position in a stream it is said to act as bait for Triclads. Because of their carnivorous habit it is wise to have a separate container for these animals when collecting.

All Triclads lay their eggs in cocoons which are often attached to water-weeds or stones. Before the cocoon is laid it can be seen in the body of the animal as a hard, roundish object, which causes the animal to bulge in about the middle or last third of the body. Those of *Polycelis nigra* are bright orange-yellow when first laid, but they rapidly darken to a dull red. Inside the cocoons are a number of eggs with a supply of food in the form of yolk ; the developing animals feed on the yolk and hatch as small creatures like their parents except in size. Some species lay cocoons all the year round, others only in winter or summer.

All Turbellaria have a great power of regeneration, that is, they can grow a new part to their bodies if it is accidentally damaged. If a specimen is cut up into several pieces each piece will grow into a new animal. Some species (e.g.

Dendrocœlum lacteum, Fig. 6B) habitually tear themselves in two by a ragged transverse line across the middle of their body. This provides a quick method of multiplying.

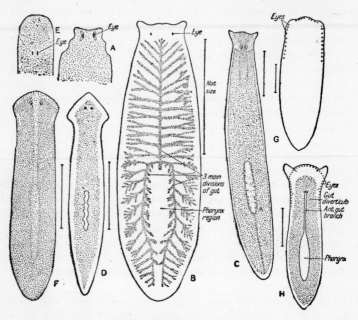

FIG. 6.—Triclad Flatworms.

A. *Bdellocephala punctata*, head end.
B. *Dendrocœlum lacteum*.
C. *Planaria alpina*, extended.
D. *Planaria gonocephala*.   After Steinman and Bresslau.
E. *Planaria vitta*, head end.
F. *Planaria lugubris*.   After Steinman and Bresslau.
G. *Polycelis nigra*.
H. *Polycelis cornuta*.

BRITISH SPECIES OF PLANARIANS (Turbellaria, Tricladida)

Bdellocephala punctata (Fig. 6A)

COLOUR.   Brown with darker spots.
EYES.   One pair near anterior end, placed wide apart.

SHAPE.   Head end with irregular anterior border, and
              distinctly marked off from the rest of the
              body when the animal is extended. Sides
              of body nearly parallel.

SIZE.   Length up to 4 cm. when extended, breadth
              about ⅔ cm. This is the largest British
              Planarian.

BREEDING   Egg capsule spherical containing 9-24
HABITS.   young.

OCCURRENCE.   Found in slow-moving or standing water,
              often in the mud. Common under stones
              in mountain tarns in the Lake District.

Dendrocœlum lacteum (Fig. 6B)

COLOUR.   Body white. Branching gut easily visible
              through body wall ; gut may be greyish-
              black or pinkish in colour.

EYES.   One pair placed wide apart.

SHAPE.   Sides of head end produced into very short
              tentacles ; sometimes these are not apparent
              and head end is then nearly square. Small
              cleft in middle of anterior border.

SIZE.   Length up to 2½ cm., breadth about ½ cm.
              when extended.

BREEDING   Spherical cocoons containing 5-42 young.
HABITS.   Breeds all the year round.

OCCURRENCE.   In streams and standing water. Very common.

Planaria alpina (Fig. 6C)

COLOUR.   Usually greyish, but white specimens or almost
              black are also found. The branching gut
              is easily visible in the grey or white specimens,
              and it may have a yellowish or pinkish
              colour.

EYES.   One pair slightly kidney shaped. In grey
              and black specimens they are surrounded
              by a whitish area, but this does not show
              in the white variety.

SHAPE.   Body slender when extended with usually
                small but well-marked tentacles at the head
                end. Posterior end pointed, occasionally
                rounded which may be due to recent trans-
                verse division. Very active when dis-
                turbed.

SIZE.   Length about 1-1½ cm., ½ cm. broad.

BREEDING   Spherical cocoons unattached to submerged
    HABITS.   objects containing 15-30 young. Cocoons
                are laid all the year round in mountain
                regions, but only during the winter and early
                spring in other places.

OCCURRENCE.   Found in mountain brooks and springs or
                small lakes. Lives at lower levels during
                the winter when the temperature of the
                water falls. Creeps upstream as weather
                becomes warmer.

Planaria gonocephala (Fig. 6D)

COLOUR.   Brown to grey, often blackish or olive-green,
                sometimes with dark longitudinal streaks.

EYES.   One pair. Slightly kidney shaped and sur-
                rounded by a whitish patch. Placed near
                anterior end of body.

SHAPE.   Head triangular, posterior part being broader
                than the rest of the body. Body narrow,
                tapering to a point at the tail.

SIZE.   Length 2-2½ cm. Breadth about ½ cm.

BREEDING   Spherical cocoons fixed to leaves or stones,
    HABITS.   laid mainly in late spring and summer.

OCCURRENCE.   Only found in the south of England. Prob-
                ably this species has been introduced
                recently and is not a native of Britain.

Planaria vitta (Fig. 6E)

COLOUR.   Pure white.

EYES.   One pair of small eyes placed close together
                some distance from anterior end.

SHAPE. Long and narrow, the sides of the body being almost parallel for the whole of their length. Anterior end rather variable, usually rounded, sometimes with a slight prominence in the centre.

SIZE. About 1 cm. in length, $\frac{1}{8}$ cm. broad.

BREEDING HABITS. Appears to reproduce sexually very seldom. Possesses the power of dividing its body to a marked degree, regeneration taking place very quickly.

OCCURRENCE. Lives in mud, usually underground, preferring wells and springs with a slight current of water through them. Found above ground when washed out by heavy rain.

Planaria lugubris (Fig. 6F)

COLOUR. Grey-brown to black, with a pair of lateral pigmentless streaks behind the eyes.

EYES. One pair near anterior end, each with a small white area round it.

SHAPE. Fairly broad, with a pointed anterior end, body broadest on a level close behind the eyes. Posterior end rounded.

SIZE. Length 2 cm. Breadth about $\frac{1}{2}$ cm.

BREEDING HABITS. Possesses the ability to divide itself in two, regeneration taking place quickly.

OCCURRENCE. It is found in gently running streams and in still water.

Polycelis nigra (Fig. 6G)

COLOUR. Variable. Light or dark brown, grey, black, greenish or sometimes nearly white.

EYES. A large number of eye-spots (30 or more) arranged round the front border of the body and extending about a third of the way down the sides of the body. These are difficult to see in very black specimens.

SHAPE.	Front part of the body the broadest. Anterior margin with three rounded prominences often ill-defined.
SIZE.	Length about 1 cm. Breadth about $\frac{1}{6}$ cm.
BREEDING HABITS.	Cocoons lemon-shaped or spherical, size about $\frac{1}{10}$ cm. Bright yellow when first laid, turning dark red later, they are deposited on leaves or stones chiefly in spring. These planarians breed readily in captivity.
OCCURRENCE.	Lives in standing water, or streams where there is only a little current. Very common in many places, particularly in the Broads.

Polycelis cornuta * (Fig. 6H)

COLOUR.	Variable. Usually some shade of brown, but grey, black, reddish, and colourless forms are found.
EYES.	There are a large number of eye-spots as in *P. nigra*.
SHAPE.	Very well-marked tentacles at the sides of the head end. Body pointed behind.
SIZE.	Length nearly 2 cm. when animal is extended.
BREEDING HABITS.	Cocoons are not attached to solid objects.
OCCURRENCE.	Lives in small streams originating from clear springs, and is to be found under stones in fairly swift currents. Common in some places.

Rhabdocœles.—The Rhabdocœle Turbellarians are more difficult to find on account of their smaller size. They may be distinguished from Triclads by the appearance of the gut, which is a simple sac with side pouches in some cases. It is never a complicated branching organ. In general shape they resemble the Triclads, but most genera are not more than 2 mm. long when adult, so they are not easy to see unless pond material is examined under a microscope. They move by means of a covering of cilia, but instead of only gliding along fairly slowly, the lashing cilia can cause many of these

---

* Also *Polycelis tenuis*. This further species has been recently described. It is very similar in appearance to *P. nigra*, but may be distinguished from it by the shape of the male reproductive organs.

smaller-bodied creatures to progress rapidly through the water. The body is usually transparent so that all the internal organs are visible. Some genera are dense white, but many are brightly coloured and are very fascinating to watch.

Rhabdocœles are mainly to be found in standing water ; only a few kinds live in streams. The adults are commonly not obtainable all the year round, but only during the warmer months. These animals may winter as eggs which do not complete their development until the temperature rises. Eggs are also a means of dispersal for the species as they may be carried about by the wind or on the feet of birds to new ponds. In summer rapid multiplication takes place, and a small sample of water may contain some hundreds of individuals. If water containing mosses and other plants is collected in winter and is kept in a warm room for some weeks it will often be found to contain Rhabdocœles which were certainly not apparent at first.

The internal organs show more variation than they do in the Triclads. Some genera have their mouths close to the anterior end, in others it is nearer the middle or hind end of the animal. Some have probosces at the anterior end, others have a muscular pharynx near the mouth, some have both. A pair of eyes is often present which are connected to a mass of nervous tissue called the brain (see Fig. 8B). Some sensory function is performed by ciliated pits, and also on some genera by stiff hairs on the surface of the body (see Fig. 9A). Rhabdites are easy to see in many species.

Rhabdocœles are generally carnivorous, feeding on small crustaceans such as water-fleas, but plants such as algæ and diatoms are also eaten ; one genus, *Microstomum*, is parasitic on Hydra. The sting cells of the Hydra become functional as defence organs in the body of the *Microstomum*.

Rhabdocœles lay yolky eggs which presumably contain several developing young. These may pass out of the adult's body, or they may be liberated only after the death of the parent. Reproduction also takes place by means of " budding " in some genera. New individuals are budded off behind the parent, and these remain attached for some time so that strings of four or five individuals are found attached together (see Fig. 9B).

I have most often found Rhabdocœles by examining detritus and mud from the bottom of ponds, or masses of green and blue-green algæ from places where these plants are covered by spray from a stream without their being in a strong current of water.

BRITISH GENERA OF RHABDOCŒLE FLATWORMS
(Not necessarily a complete list)

Genus Stenostomum.—Several species belong to this genus. They commonly reproduce by "budding," and from two to four individuals are often found attached together (see

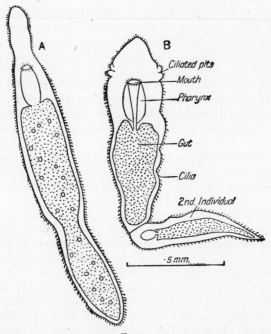

FIG. 7.

A. *Stenostomum*, swimming.

B. *Stenostomum*, chain of two individuals, slightly compressed.

Fig. 7). Colour transparent white, green, or pale brown. Near the head end are ciliated pits. The mouth is in the midline close behind the pits. The gut is a simple sac. Chains of two individuals may only measure just over a millimetre in length. Common in detritus from ponds all the year round, also among liverworts and algæ from horse troughs and other places where there is a gentle current of water. Food consists of small animals such as young ostracods and diatom plants.

Genus Prorhynchus.—Two species occur in Britain; both are common, at least in the Sheffield district.

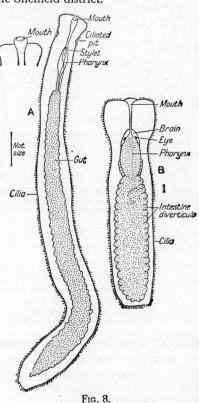

Prorhynchus stagnalis. —A long thin Rhabdocœle (see Fig. 8A), glistening white, opaque, with sluggish movements. Length up to 12 mm. The mouth, pharynx, and gut are the only structures which can be easily made out in living specimens. There is a chitinous stylet near the anterior end and a pair of ciliated pits as in *Stenostomum*. Common among mats of blue-green algæ which are in a spray of water.

Prorhynchus applanatus. —A much smaller species, only about 2 mm. long (see Fig. 8B). The body is much flattened and resembles that of a Triclad; it is usually pale brownish. Two small eyes are

FIG. 8.
A. *Prorhynchus stagnalis.*
B. *Prorhynchus applanatus.*

present, situated above the nervous mass or brain. The mouth is at the anterior end and it leads into a very muscular pharynx.

Genus Macrostomum.—Probably two species in Britain. Individuals are rather small, less than 2 mm. long. The anterior end has long sensory hairs projecting from it (see

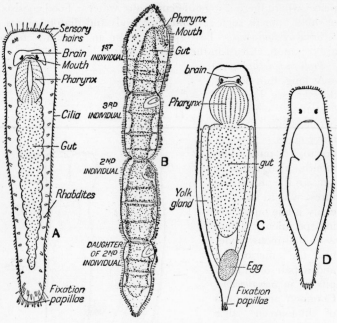

FIG. 9.

A. *Macrostomum*, immature specimen, length ·75 mm.
B. *Microstomum*, 2 mm. long. After Wesenberg-Lund.
C. *Dalyellia*, length 1·4 mm.
D. *Dalyellia*, young specimen.

Fig. 9A), while the posterior end has large papillæ for fixing the animal to surfaces. Rhabdites are very obvious in the body. A pair of eyes and a " brain " are easily made out. The mouth is immediately behind the brain and leads into a very muscular pharynx.

Genus Microstomum.—Species of this genus are usually found

in chains of four or more individuals, which have been produced by budding. The gut extends forwards, beyond the
pharynx (Fig. 9B). They live on Hydra, and the sting
(*cnidoblast*) cells of that animal are taken into the body of
Microstomum where they migrate into the skin and become
functional, their *nematocysts* being used as defence organs.

Genus Dalyellia.—Several species, a common one being bright
green. The body is usually square or rounded in front,
tapering behind. The characteristic feature is the cask-
shaped pharynx which opens at the mouth near the anterior
end (Fig. 9C, D). A pair of kidney-shaped eyes are present
above the brain. Like the genus *Macrostomum*, *Dalyellia* has
attachment papillæ at the posterior end. On each side of
the gut is a long simple yolk gland which produces the yolky
food material for the eggs. A single egg with a hard protective
covering is often found in the body of adult individuals.
This genus is common among plants in still ponds, it feeds
on diatom plants.

Genus Opistomum.—One species, probably
O. pallidum, a robust Rhabdocœle, oval in
cross-section and about 4 mm. long (Fig. 10).
Colour of body pale pink, with the gut showing greenish, some of the organs yellow,
others white, and the mature eggs when
present dark red. The mouth is near the
posterior end and leads through a muscular
pharynx into the simple gut. The anterior
end is rounded ; there are no eyes. Found
among water-weed in small pools, particularly those which are liable to dry up in
summer. Numerous resting eggs are produced inside the body (Fig. 10).

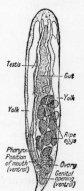

FIG. 10.
Opistomum. Dorsal
view, length 3·5
mm.

Genus Rhynchomesostoma.—One species, *R.
rostratum.* This brilliantly coloured Rhabdocœle is easily recognised by its proboscis
at the anterior end (see Fig. 11A). It is about
2 mm. long, transparent rose-pink in colour with red eyes, red
yolk gland, and dark red resting eggs when these are present.
The mouth and pharynx are near the middle of the body.

This Rhabdocœle is very common in small pools and is found along with *Opistomum*. Individuals in December and January contain resting eggs which are set free with the death of the parent (Fig. 11B).

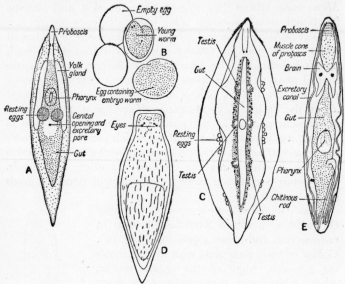

FIG. 11.
A. *Rhynchomesostoma rostratum*, length 2 mm.
B. A group of resting eggs of *Rhynchomesostoma rostratum*.
C. *Mesostoma tetragonum.* After Braun.
D. *Castrella*.
E. *Gyratrix*, length 7 mm.

Genus Mesostoma.—Two species in Britain. Members of this genus are flat and leaf-like in shape, very transparent and often fairly large (up to 14 mm. long) (Fig. 11C). Colour pale brownish with some of the organs yellow, resting eggs dark red. These animals produce two kinds of eggs ; those of one type develop quickly and hatch inside the parent ; they are produced under favourable conditions. The resting eggs, on the other hand, have thick shells, they are resistant to unfavourable conditions and can survive long periods. They only escape with the death of the parent.

Genus Castrella.—A small brown species, which seems to be fairly common among floating water-plants, especially inside hollow stems. The brown pigment is in streaks which obscure the internal organs in living specimens. The eyes are rather characteristic, as they are each almost divided into two parts which remain connected only by a narrow bridge (Fig. 11D).

Genus Gyratrix.—One species, *G. hermaphroditus.*—The special characteristics of this genus are first a papillated conical proboscis at the anterior end which does not protrude from the body, and secondly a long chitinous structure near the posterior end (see Fig. 11E). Specimens are about 2 mm. long, white, and very transparent with black eyes. The mouth and pharynx are in the middle of the body. A pair of long coiled excretory tubes are easy to see on either side of the animal.

For further reference to Turbellarians (Triclads and Rhabdocœles) see

1. Die Strudelwürmer (Turbellaria), von Dr. P. Steinmann und Dr. E. Bresslau. Leipzig, 1913.
2. Fresh-water Biology, by H. B. Ward and G. C. Whipple. New York, 1918.
3. British Species of Polycelis (Platyhelminthes), Reynoldson, T. B. (1948), *Nature*, **162,** 620.

(*b*) **Class Trematoda.**—The Flukes (parasitic for the greater part of their lives).

Most adult forms of Trematode flatworms are similar in shape to the free-living Triclads, being flattened and leaf-like. Because they are parasites and have to cling to their host they have suckers on the underside of their bodies ; some have two suckers, others have more. The mouth is always near the anterior end, and is in the middle of the depression of the front sucker. The mouth leads through a muscular pharynx into the gut which branches much like the gut of a Triclad. The adult body is not covered by cilia but has instead a thick cuticle on the outside. There are no eyes in the adults of these flatworms, and the brain is very small. This is probably because adult flukes do not have to search for their food once they have found a suitable host on which to become a parasite. As is common in parasitic animals an enormous number of eggs are produced, only a very few of which ever develop as far as the adult condition, through failure to find the correct host.

There are two types of Trematode flatworms. In the first the life-history is simple, and such species only parasitise one host. In the second type two or more hosts are involved making the life-history complicated. Those forms which

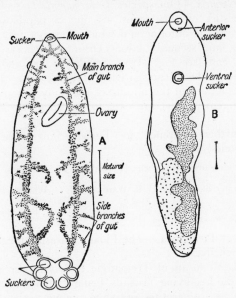

FIG. 12.—Trematode Flatworms.
A. *Polystomum*, adult. From a frog's bladder.
B. *Pneumonoeces*, a Trematode parasite, found in frogs' lungs.

only attack one host are found as external parasites on the gills of fishes, but they are rare in fresh-water, being more common on sea fishes. Closely allied to these fish parasites are some forms which are found on the body or in the alimentary canal of Reptiles and Amphibia. The common frog will be found to have one such flatworm, called *Polystomum* (Fig. 12A), in the bladder of between 3 and 5 per cent. of frogs examined. When frogs are being dissected it is always worth

while to open up the bladder of all the specimens in the chance
of finding it. The same Trematode is also found on the
external gills of frog tadpoles. The life-history of *Polystomum*
(see Fig. 12A) and other Trematodes which are external parasites, is as follows : The eggs, which are laid on the host (or
pass into water as is the case with *Polystomum*), hatch as
ciliated larvæ with eye-spots and a ventral sucker. These
larvæ have a free existence for a short time, but as soon as
they find a new host they attach themselves to it, loose their
ciliated covering and eye-spots, and grow into an adult.
The larva of *Polystomum* attaches itself to the gills of a tadpole
after a free existence of not longer than twenty-four hours.
From the gills it wanders into the bladder, but it does not
become adult for three years. If a *Polystomum* larva finds a
tadpole which is young and has external gills, it attaches
itself to these. The Trematode then becomes adult very
quickly, but it dies when the tadpole metamorphoses into
a frog.

The second type of Trematodes are all parasites of the
internal organs of their host, such as the liver or brain. Their
life-histories are complicated and involve two or more hosts.
In some cases all the hosts are fresh-water animals, but often
the final host in which the flatworm becomes adult is a land
mammal or bird ; one of the subsidiary hosts is always a
mollusc. Three examples will give you some ideas about
these animals.

Sheep in great Britain sometimes suffer from a disease
called " Liver Rot." This is due to flukes being present in
the bile ducts of these animals which is the normal place for
these flukes to live as adults. The eggs laid by them pass
into the intestine of the sheep with the bile from the liver
and eventually reach the outside. They are unharmed by
the digestive process of the sheep, but they do not develop
unless they reach damp grass, during warm weather. After
a few weeks the eggs hatch into a Miracidium larva
(see Fig. 13A) ; this larva swims with its coat of cilia until it
meets a certain water-snail (*Limnaea truncatula*, see p. 239,
sometimes *Limnaea pereger* is attacked if there is no *Limnaea
truncatula* available), or if it does not meet such a snail it dies.

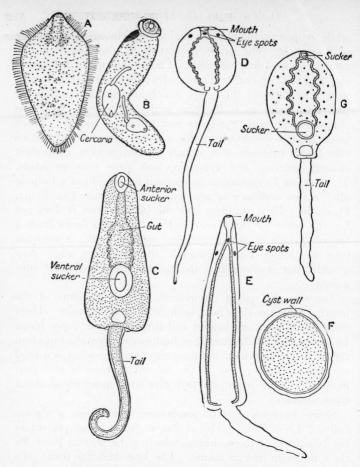

FIG. 13.—Young stages of Trematode Flatworms.

A. Miracidium larva.
B. Redia stage, showing developing Cercariæ.
C. Cercaria, obtained from the snail *Limnæa truncatula*.
D. Cercaria monostomi, swimming, from the wandering snail *Limnæa pereger*.
E. Same, creeping.
F. Same, encysted.
G. Cercaria granulosa.

Those larvæ which find the right snail bore their way into it and undergo extensive development inside (see Fig. 13B). The Miracidium larva grows into a large sac and within the sac (sporocyst) many new larvæ are formed which are different in shape from the first ones. They are called *Rediæ*; after living on the snail and sometimes killing it they migrate inside the snail, and give birth to *Cercariæ* larvæ with long tails (see Fig. 13B-E, G). These come out into the surrounding water and swim about for a time. The normal snail host lives in marshy meadows, not just in permanent ponds so that the Cercariæ may find themselves in shallow water in which grass is growing. The Cercariæ settle on blades of grass and secrete round themselves a protective covering (Fig. 13F). Nothing further happens unless such blades of grass are eaten by a sheep. If they are, then the cyst is dissolved in the sheep's gut and the larva migrates to the liver where it grows into an adult. Sheep can only become infected with flukes when they are kept on damp pasture and when there is the correct snail in the district to act as the intermediate host.

Another Trematode with a complicated life-history is *Gasterostomum* (*Bucephalus*). The adults are found in perch, The Miracidia bore their way into *Anodonta*, a fresh-water bivalve mollusc (see p. 245). The Cercaria larvæ encyst in the mouth of the roach, and if the roach is eaten by a perch, then the cysts grow into adults.

Water-snails of the genus *Planorbis* (p. 242), especially the species *Planorbis albus*, are hosts for the Miracidia and Rediæ stages of a Trematode *Cotylurus*. The Cercaria of this parasite encyst in leeches of the Herpobdella type (p. 60) which feed on the snails. One quite small leech may contain three or four lemon-shaped cysts. These do not develop any further unless the leech is eaten by a bird, probably some kind of water-fowl.

Occasionally when a frog is being dissected, its lungs are found to be distended with several black and cream-coloured objects ; these are the adults of a Trematode parasite called *Pneumonoeces* (Fig. 12B). If they are carefully removed and placed in Ringer's solution they may be seen to move slightly. The body is flattened on the ventral side, slightly convex on the dorsal. Two suckers can easily be made out. Part of the body is creamy-white, part black, and part orange or yellow.

Rediæ and Sporocysts of various Trematodes can be obtained by dissecting almost any species of water-snail and examining the liver under a microscope. The free-swimming Cercaria larvæ may often be had in numbers if several good-sized specimens of water-snails, such as *Limnaea stagnalis* or *L. pereger*, are placed in small individual tubes containing water and left for several days, after which time Cercariæ may be found swimming about or encysted on the glass. Certain times of year, such as the autumn and the early summer, seem to be better than others for getting these larvæ. They are well worth examining alive because of their fantastic movements ; one second they are swimming violently with their tails greatly elongated while the body is more or less rounded, the next they are creeping along the glass surface, their bodies elongated and their tails short and fat. They often cast off their tails and encyst while you are examining them. After being shed the tail goes on wriggling for some time. Some Cercariæ have eye-spots (Fig. 13) while others have none ; two suckers are generally present, but there may be only one in a few species. One type has a forked tail (see Fig. 13 for various Cercariæ).

Class Cestoda.—The Tapeworms

A typical adult tapeworm has a long thin and narrow body which is not transparent. As adults they are found living as parasites in the alimentary canal of vertebrate hosts. The worm is narrower at the head end than it is at the tail. During life the head (*scolex*) is buried in the wall of the host's gut, it being provided with suckers and sometimes hooks as well for keeping a hold on the host. Tapeworms have no gut and no mouth. They obtain their food by absorbing digested material through their surface from their host's alimentary canal. The ribbon-shaped body is divided transversely into a large number of compartments (*proglottids*) each with a complete set of reproductive organs which produce very large numbers of eggs. New compartments

are continually being added from near the head region so
that the worm grows longer and its capacity for egg-produc-
ing increases. In some forms the compartments near the
tail drop off when the eggs are ripe, in others the eggs escape
through a pore. In either case the ripe eggs pass to the
exterior with the undigested food of the host. To complete
the life-history one or two intermediate hosts are required;
in many cases the intermediate hosts are not known. As
with the Trematodes all the hosts may be fresh-water animals, but commonly one is a mammal or a bird. There
are also many Cestodes for which the hosts are all land
animals.

When all the hosts of Cestodes are fresh-water animals
then the adult worms live in the alimentary canals of fish.
The eggs hatch in the surrounding water into a larva with
a ciliated coat. This larva which has a short free existence
is provided with hooks, and it bores its way into some small
fresh-water animal such as a *Cyclops* (see p. 97), an Ostracod
(see p. 87), a small Annelid worm, or more rarely a snail.
Here it forms a cyst in the host's tissues and does not develop
further unless its host is eaten by a larger animal such as a
small fish. This fish may not be the final host, in which case
the tapeworm grows inside its new host and then forms a
much larger cyst. This second cyst develops into the adult
only when the fish is eaten by a larger fish. Sticklebacks are
sometimes found which have a large bulge on one side of the
body. This bulge is the second cyst of a tapeworm called
Schistocephalus gasterostei; in this species the worm becomes
adult when the stickleback is eaten by a bird. (This should
be distinguished from the white nodules produced on
sticklebacks by the protozoan parasite *Glugea*, see p. 286.)

For further reference to *Trematoda* and *Cestoda* see

Fresh-water Biology, by H. B. Ward and G. C. Whipple.
    New York, 1918.

(c) Note on *Nemertean* Worms

The Nemerteans are a small group of animals with some
resemblances to the free-living flatworms. They are covered

with cilia and have an eversible proboscis. Most species are marine, a few are terrestrial, and two or three are inhabitants of fresh water. One species belonging to the genus *Tetrastemma* (*Emea*) has been found in Britain, but it appears to be more common in France and Switzerland.

Tetrastemma is about 2-3 cm. long with a flattened ciliated body which is narrower than those of the Triclad Turbellarians, being only about 1 mm. wide. Near the head end are six eye-spots arranged in two rows. The body is transparent, and either brown, orange, or reddish in colour. The animal is found among mud at the edge of lakes and slow-flowing rivers. When moving it glides like a flatworm ; it does not swim, and is often found curled up. The mouth is very close to the anterior end, and through this opening the animal can protrude a long thin retractile proboscis. This proboscis lies when not extended in a special long sac which extends most of the way down the body on top of and separate from the gut. This structure is one of the chief characteristics of Nemerteans ; it can be seen through the transparent body of *Tetrastemma*, and together with the animal's narrow body and anteriorly placed mouth, serves to distinguish it from the flat-worms. The eggs are laid on water-plants in jelly masses.

CHAPTER 5

THE ROUNDWORMS (Nematoda)

THIS group of animals contains a large number of " worms "
with long cylindrical bodies, spindle shaped or sometimes
thread-like. The body is usually pointed at both ends and it
shows no sign of segmentation such as is characteristic of the
true worms. On the outside is a clear elastic cuticle which is
shed four times during the period of development from egg to
adult. Near the anterior end is a mouth (which is lacking in
some forms). This leads through an enlarged fore-gut often
with a bulbous portion (Fig. 14) to the main part of the alimentary canal, which is a straight tube leading to the anus at
the posterior end of the body. The mouth is sometimes armed
with hooks and stylets. Individuals are of separate sexes developing either sperms or eggs in a large mass of cells alongside
the gut. The female has a special exit for the eggs, but in the
males the sperms are passed out through the anus. A curious
feature of the Nematodes is that the tissues of which their
bodies are composed are made up of very few large cells ;
there are no cilia anywhere, no blood system, and no breathing
organs. The eggs are covered by a hard shell. The young
Nematode hatches looking very much like an adult except
for size. Sometimes they hatch inside the body of the female,
but they generally do so after the eggs are laid. The young
may encyst inside a shed skin, in which state they can
withstand desiccation.

Many of the Nematodes are parasites for at least part of
their lives. A large number attack plants, and many are
internal parasites of other animals. A number of free-living
genera are common in fresh water particularly among mud,
but they are small, the average size being 1 mm., and they

43

are not easy to identify. The movement of a Nematode is quite characteristic; they progress through mud or water by contracting the body into S-shaped curves; they have no circular muscle in their body wall so they are not able to move like a true worm by alternately stretching their bodies so that their diameter is decreased then contracting them so that the diameter increases. Only two common genera are described here because a good deal of experience is needed with the group before they can be easily identified.

Genus Rhabdolaimus.—There is one very common species of this genus, *R. aquaticus*, which is a small transparent Nematode about 1 mm. long (Fig. 14). It has no hooks or stylet at the mouth end. It is found among pond mud and débris.

Genus Dorylaimus.—Several species are found in mud at the bottom of ponds and among the roots of water-plants. One common one is large for a free-living

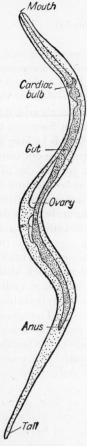

FIG. 14.
Rhabdolaimus.

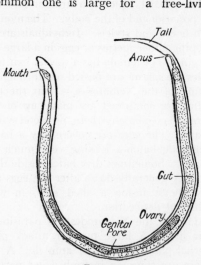

FIG. 15.
Dorylaimus stagnalis, female.

Nematode being 5-8 mm. long. It is *Dorylaimus stagnalis* (Fig. 15). The mouth has a stylet.

There are very many more fresh water Nematodes, some of them living as parasites inside Annelid worms. When you are examining pond material (mud, vegetable débris, etc.) under the microscope you will come across them constantly, and you should be able to recognise them on sight because of their unsegmented bodies and S-shaped curves of the body when they move.

Book for further reference—

Brauer, A. Die Süsswasserfauna Deutschlands, Heft 13 : Nematodes. Jena, 1909.

Goodey, T. Soil and Freshwater Nematodes. London, 1951.

Note on *Gordiaceæ*

The Gordiaceæ are a small group of worms which in many ways are like Nematodes. There is only one fresh-water genus, *Gordius*, the species of which are commonly called " Hairworms." The adults are anything from 11 to 65 cm. long, the body is opaque and exceedingly narrow with a forked end in the male. They are found curled up, sometimes several individuals entwined round each other, in any kind of stagnant water. The animals are of separate sexes, and in the adult stage they apparently do not feed, their sole function being to reproduce the species. The females lay long strings of eggs attached to plants. Very little is known about the life-history, but the larva which hatches from the egg has a boring apparatus at the head end with which it bores its way into the body of some insect larva such as that of a Dytiscus beetle, a Mayfly, or an Alderfly. Inside the body cavity it leads a parasitic life growing into a long thread-like creature. At some stage it leaves its host, becomes adult, and leads a free (non-parasitic) existence.

Note on *Acanthocephala*

The group of animals called Acanthocephala or Proboscis Roundworms are also thought to come somewhere near the Nematodes in their affinities. They are parasites showing much modification for this kind of existence ; for example, they have no alimentary canal, and absorb food from their host through their skin. They have at the anterior end a stout proboscis armed with rows of recurved hooks, which can be retracted inside a sheath. All the fresh-water species are small being only 2 or 3 mm. long.

There are several species of *Echinorhynchus*, the young stages of which are found in Crustacea while the adults occur in fish. The fresh-water shrimp (*Gammarus pulex*) is the first host for one species ; here the parasite lives in the body cavity and is visible through the animal's transparent skin. The larvæ only become adults if the shrimp is eaten by a fish when they attach themselves to the wall of the intestines. Another species has its larval stage in the fresh-water louse (*Asellus*), the adults also being found in fish.

CHAPTER 6

THE TRUE WORMS (Annelida)

THE true worms have soft segmented bodies often more or less cylindrical in shape. The skin is covered by a layer of cuticle (a non-cellular substance secreted by the skin cells underneath), which is soft and is not made hard by the inclusion in it of other substances as is done in the Insects and Crustacea. The segmentation of the body is marked on the outside by a series of rings which extend all the way down at approximately equal distances apart. All or some of the rings on the outside correspond with divisions inside the worm which result in the internal part of the worm being sectioned off into self-contained compartments. The organs inside each compartment are largely the same, and this is the type of segmentation called *metameric*. The internal divisions are known as septa, and they can be seen easily through the transparent bodies of some of the water-worms. Many worms possess bristles (*setæ*) on their bodies which are arranged in groups on each segment ; they are used to obtain a hold on the sides of burrows, and they are a very valuable means of identifying a specimen. Worms all have distinct heads and tails though it is sometimes difficult to see which is which, but the mouth is overhung by a small lobe whereas the anus is terminal. Worms are found in the sea, in the soil, and in fresh water ; most are free living, though many construct burrows ; only one small group the Leeches, has adopted the habit of living on the blood of other animals.

The general arrangement of internal organs in the fresh-water worms is similar to that of the ordinary earthworm which you may have dissected. In many of the transparent forms you can easily observe the movements of the gut, blood-vessels and *nephridia* (paired excretory organs occurring in most segments) in the living creature, which will give you a good idea of what is really going on inside the more opaque body of a living earthworm. All these worms have both male and female reproductive organs in the same animal, so that every specimen lays eggs when adult. Like the earth-worms these eggs are placed in a cocoon secreted by a girdle of special cells called the clitellum, which shows as a swollen area of skin near the middle of the body. In addition to this method of reproduction by laying eggs, several families of water-worms have an asexual method of multiplying by " budding " off daughter worms near the hind end of the body. The daughter worms eventually separate off from the parent, but before this occurs several worms may have been budded, forming a chain of individuals.

The worms found in fresh water belong to two groups, of which one possesses bundles of bristles on most segments (the Chætopoda) while the other do not (the Leeches or Hirudineæ). These two groups have rather different habits and they will be treated separately.

(a) Fresh-water worms with bristles (Chætopoda Oligochæta)

This group of water-worms are more like the earthworms than are the leeches. A number of them construct burrows in the mud at the bottom of ponds. The burrow is often prolonged above the surface of the mud as a tube made by the worm from particles of debris. The worms live inside the tubes with their heads at the bottom of the burrow feeding on the mud while their tails wave about in the water. They are very sensitive to touch and will retract rapidly into their burrow if they are touched or even if the water round them is disturbed. If you wish to collect tube-building worms you must scoop up a quantity of mud containing them and place it in a jar with water above. In a day or two the worms in the mud will have constructed new tubes and you can then observe their habits.

The fresh-water worms with bristles belong to eight different families which are fairly easy to distinguish from one another, but in some of the families it is very difficult to identify the various genera without making sections of the animals. In the short description of the families given below those which include a number of common easily identified genera are treated more fully than the others. In nearly every case the bristles (*setæ*) are the important structures to examine. For an examination of the bristles the worm should be placed in some water between a slide and cover-glass and examined under the microscope. The bristles are always arranged in four bundles on each segment, two bundles on the underside (ventral) and two more at the side near the upper surface (dorsal setæ). In some worms they are absent from several segments. The bristles of the underside are often of quite a different size and shape from those above, and when they are small you may find it quite difficult to make out whether they are cleft or not at the tip. It may be necessary to kill the worm and squash part of it flat under the cover-slip in order to see the bristles better.

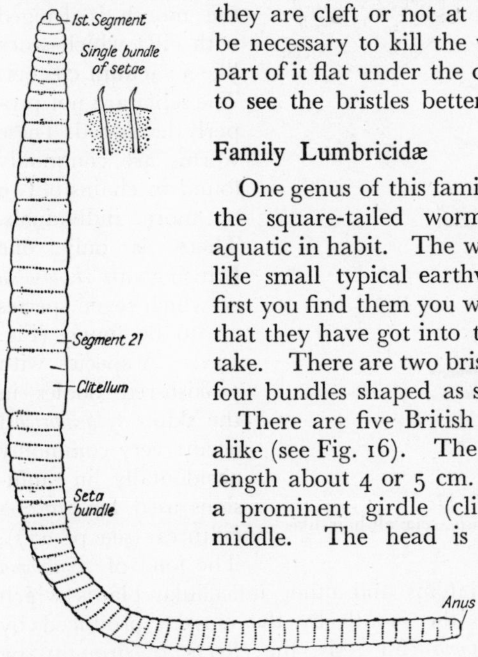

Family Lumbricidæ

One genus of this family of earthworms, the square-tailed worm (*Eiseniella*), is aquatic in habit. The worms look exactly like small typical earthworms, and when first you find them you will probably think that they have got into the water by mistake. There are two bristles in each of the four bundles shaped as shown in Fig. 16.

There are five British species all much alike (see Fig. 16). The colour is pinkish, length about 4 or 5 cm. and the body has a prominent girdle (clitellum) near the middle. The head is slightly pointed while the tail ends squarely. These worms are found in the beds of mountain streams

Fig. 16.—The Square-tailed Worm (*Eiseniella*).

4

or among moss and algæ in situations where these are constantly sprayed with water, such as an old disused waterwheel or the overflow from a mill lade. They also occur among the mud at the edge of lakes. In such places they are common.

Family Æolosomatidæ

Small transparent worms 1-2 mm. long with all the bristles hair-like. The skin contains conspicuous rounded bodies coloured pink, yellow, or green depending on the species. The under surface of the head is covered with cilia and the mouth is fringed with cilia which work like a vacuum cleaner. The septa are not properly developed. These worms are commonly found in chains of two or more individuals. There is only one British genus *Æolosoma* of which seven species are to be found (Fig. 17A). A species with pinkish-red bodies in the skin (*A. hemprichi*) occurs very commonly accidentally in infusions used for amœba cultures (see p. 277). The food of *Æolosoma*

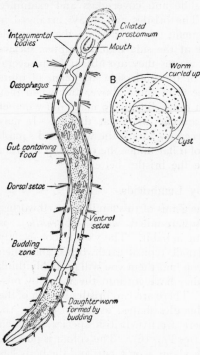

FIG. 17.
A. *Æolosoma hemprichi*, setæ all hair-like.
B. *Æolosoma* cyst.

appears to be diatoms and other unicellular plants which are whisked into the mouth by the current produced by the cilia. *Æolosoma* can resist the effects of drought by enclosing itself in a cyst and remaining dormant (see Fig. 17B).

Family Naididæ

This is a large family all the members of which have the bristles in the ventral bundles cleft at the ends, as shown in Fig. 18A. Most of the genera belonging to the family are found in chains of several individuals, and the body colour is usually white or pinkish. Reproduction by budding is much more common than by laying eggs. The size varies from 2 or 3 mm. up to 20 mm., but it is seldom more. There are eleven genera with thirty species between them in this country.

Genus Chætogaster (Fig. 18A).—Commonly found associated with tube-building insect larvæ such as Chironomus (see p. 210) or inside the shells of the wandering snail (*Limnæa pereger*). The worm is whitish except for the gut, and there are no bristles on segments 3, 4, and 5. The anterior part of the body is very broad. Five species.

Genus Nais (Fig. 18B).—A very common worm found amongst detritus at the bottom of ponds where it builds rather flimsy tubes. Colour often pale pinkish or brownish with usually a pair of eye-spots. Dorsal bristles long and hair-like, ventral small and cleft. Frequently found in chains ; size up to 25 mm. Four British species.

Genus Stylaria (Fig. 18c).—This genus is very easily recognised because the front end of the worm is prolonged into a very distinct long narrow proboscis. There is a pair of eye-spots. Two species commonly found among floating water-plants in ponds and the Norfolk Broads.

Genus Dero (Fig. 18D).—Members of this genus live in mud and construct tubes. They have red blood which makes the body look pink. At the posterior end there are a number of ciliated " gills " in which the red blood may be seen circulating. The gill area can be expanded or retracted. There are seven species.

Genus Paranais (Fig. 18E).—This worm is pure white and very transparent. At rest the body is held straight except for a gentle curve at the anterior end, if disturbed the worm becomes very active. All the bristles are short and cleft at the end. There are two species in Britain found among floating weeds in ponds and lakes.

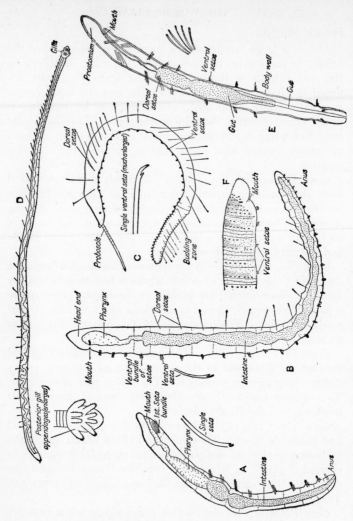

Fig. 18.—Oligochæte worms, family *Naididæ*.

A. *Chætogaster*, length 7 mm.
B. *Nais*, side view, length 15 mm.
C. *Stylaria lacustris*, length 15 mm.
D. *Dero*, whole worm and tail end more highly magnified, length 19 mm.
E. *Paranais*, length 1·5 mm.
F. *Ophidonais*, head end, total length 30 mm.

Genus Ophidonais.—These are among the largest worms in the family Naididæ, reaching a length of 3 cm. The bristles in both dorsal and ventral bundles are short and a pair of eye-spots are present at the head end. There are two species in Britain of which *O. serpentina* has four bands of pigment near the anterior end (Fig. 18F). They are often found in chains of several individuals. They live in mud at the bottom of ponds and lakes.

Genus Pristina.—In this genus the dorsal setæ are hair-like and they have very fine teeth down one side. The head end of the worm is often produced into a proboscis. There are three British species.

There are four other genera of Naid worms which are less common. They are *Naidium*, *Vejdovskyella*, *Slavina* each with one species, and *Ripistes* with two species.

Family Tubificidæ

Members of this family have usually red blood which makes

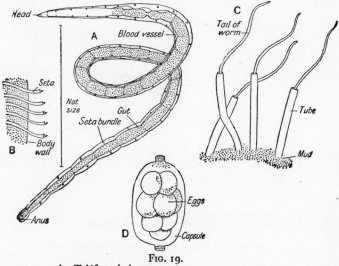

FIG. 19.

A. *Tubifex*, whole worm.
B. Single bundle of setæ.
C. *Tubifex* worms in mud tubes.
D. *Tubifex* egg capsule. After Wesenberg-Lund.

the blood-vessels stand out very clearly in the transparent
white body.   The bundles of bristles have more than two
short bristles each (Fig. 19B), and the ventral ones are always
hooked with the tips cleft.   The worms are usually more than
3 cm. long (Fig. 19A).   They are found in mud under water,
and they construct tubes out of which their tails project and
wave about (Fig. 19C).   The mud is passed through their
bodies, and any digestible material is removed on the way.
The tail is used as a gill to obtain oxygen from the water as
there is very little of this gas in the mud where the main part
of the worm is buried.   In water which is poorly oxygenated
a greater length of tail is extended into the water than when
it contains more oxygen.   The eggs are laid in characteristic
capsules (Fig. 19D).   There are six genera all very difficult to
identify ;   *Tubifex* (see Fig. 19A) is one of the most common.

Family Lumbriculidæ

Long thin worms measuring up to 8 cm. in length appearing
dark red owing to the colour of the blood.   The dorsal blood-

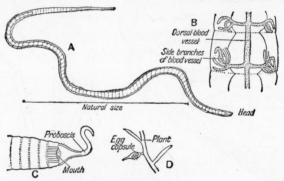

FIG. 20.
A. *Lumbriculus variegatus*, whole worm.
B. *Lumbriculus variegatus*, two segments enlarged.
C. Head end of *Rhynchelmis*.
D. Egg capsule of *Rhynchelmis*.   After Wesenberg-Lund.

vessel gives off a number of blind branches in the mid-body
region (Fig. 20B).   Only two setæ in each bundle (making
eight setæ in all) which may be cleft or simple pointed.   There

are four genera of which *Lumbriculus*, containing one British species *L. variegatus*, is a very common example (see Fig. 20A). It is found everywhere living in tubes among the mud along the edges of ditches, streams, rivers, lakes, etc. The body is greenish at the anterior end, with the deep red colour of the blood showing through wherever it is not obscured by green pigment. Apparently it never reproduces by laying eggs in this country ; multiplication takes place by the worm fragmenting and each piece growing a head and tail. The three other genera have transparent bodies without pigment ; they are *Stylodrilus*, *Rhynchelmis*, and *Tricodrilus*. *Rhynchelmis* possesses a long more or less thread-like proboscis (Fig. 20C) and it lays egg capsules attached to plants (Fig. 20D).

Family Haplotaxidæ

Two British species belong to the genus *Haplotaxus*. These worms are very long and thin, 15-30 cm. long and less than a millimetre broad. The body being made up of about 480 segments (Fig. 21). They have two pairs of bristles in each bundle which are simple pointed. The ventral setæ are larger than the dorsal ; the latter may be absent from several of the anterior segments. The worms are sometimes subterranean in habit, but they are also found in marshes, ditches, and in the mud near the banks of rivers.

FIG. 21.—*Haplotaxus*, natural size 30 cm.
After Wesenberg-Lund.

Family Enchytræidæ. *The Pot Worms*

This is a large family with 10 genera and a total of 130 species found in Britain. Most of them are not more than $2\frac{1}{2}$ cm. long, whitish in colour, and with a number of simple-pointed bristles in each of four bundles which are either straight or slightly S-shaped (Fig. 22C, D). They are to be found among the roots of plants growing in water or among mosses and algæ in situations where these are kept damp by a spray of water. A clitellum is obvious in some, as for instance in *Lumbricillus* (Fig. 22A), but in others it is not so easily made out. The genus *Enchytræus* (see Fig. 22B) is very common, and members of this genus have the habit of keeping the body very rigid. On account of this they closely resemble the white roots of water-plants among which they are found.

For further reference see

1. The British Annelids, by H. Friend.

2. Fresh-water Biology, by H. B. Ward and G. C. Whipple. New York, 1918.

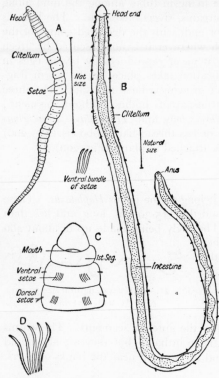

Fig. 22.

A. *Lumbricillus*, whole worm.
B. *Enchytræus*, ventral view.
C. *Lumbricillus*, head end.
D. *Lumbricillus*, single bundle of ventral setæ.

(b) Leeches (Hirudinea)

There are eleven different kinds of fresh-water Leeches in Britain, some of which are very common in ponds and streams. They are generally found attached to stones or water-weed, and at first sight they look like lumps of brown, yellow, or green jelly. If a stone with leeches attached is removed from the water and examined, the "lumps of jelly" will often be seen to glide erratically along the wet surface of the stone. Like earthworms, they have long narrow bodies (at least when they are extended), and careful examination shows that their bodies are similarly divided up into a large number of segments. They may be distinguished from all other worms by the presence of two suckers, one at either end of the under-surface of the body, and by the much greater flattening of the body (i.e. they are not cylindrical in cross-section). They also possess eyes at the head end varying in number from two to ten according to the species. The hind sucker is easily seen in all leeches (see Fig. 23), but the sucker surrounding the mouth at the head end is often only visible from below while it is extruded. These structures are used to obtain a firm hold on solid objects while the animals are at rest, and to grip surfaces when the animals are moving.

Many leeches have patterns of stripes and spots on their backs. Their bodies are often so transparent that internal organs can easily be seen through the skin. In fact one of the outstanding features of many is a dark bluntly branched tree-like organ which shows through in the middle of the back. This is part of the alimentary canal ; it is most readily seen when its owner has recently had a meal. The skin contains star-shaped pigment cells (chromatophores), the contraction or expansion of which enable the animal to lighten or darken the general body colour. The pigment cells are affected by light. In bright light they contract, making the leech lighter coloured and so more likely to match its background. In dull light the pigment cells expand, giving the reverse effect. Chromatophores of leeches do not seem to be as efficient as those of some other animals, for instance those of the frog. Colour and markings of any one species of leech vary, there are often several colour varieties. It is therefore necessary when identifying to rely on more constant features, such as the shape of the body, number and arrangement of the eyes and method of moving.

Leeches usually live by sucking the blood of other animals, only a few eat whole small animals. The mouth is situated the centre of the head sucker. Some leeches (*Rhynchobdellidæ*) pierce the skin of their host by means of a proboscis ; others (*Gnathobdellidæ*) by means of jaws. Blood-sucking leeches are external parasites of fish, frogs, water-snails, or insect larvæ in the same way as fleas and lice are external parasites of man.

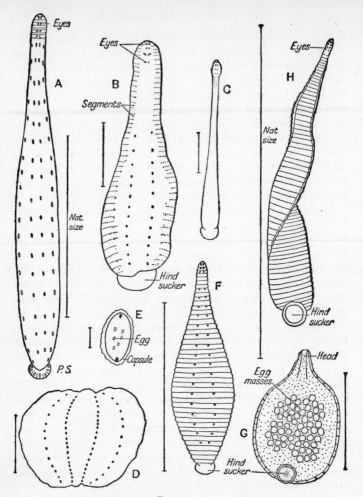

FIG. 23.
A. *Protoclepsis tesselata*, extended.
B. *Protoclepsis tesselata*, at rest.
C. *Protoclepsis tesselata*, young specimen, extended.
D. *Protoclepsis tesselata*, much contracted.
E. Egg capsule of *Herpobdella*.
F. *Glossosiphonia*, dorsal view.
G. *Glossosiphonia*, ventral view.
H. The horse leech (*Hæmopsis*).

They remain attached to their hosts until they have obtained as much blood as they can hold ; they then drop off to spend the next few weeks or months digesting and absorbing the meal.

Some leeches swim through the water by throwing their bodies into a series of undulations.   The horse leech (Fig. 23H) can often be made to do this in captivity by placing a specimen in a dish of water so that it floats ; it is then unable to get one of its suckers attached to the side of the dish without swimming. All leeches move about by using their suckers.   The hind sucker is attached to a piece of water-plant or a stone, the body is then elongated and waved about until the anterior sucker touches something solid.   The leech then grips with the anterior sucker, lets go with the hind sucker, then contracts its body so that the hind sucker can be attached to the same solid object as the front one.   This series of movements is repeated, the whole process being very similar to the method of progression seen in a " looper " caterpillar.   The suckers can obtain a very firm hold, which explains why leeches may be found in swift-flowing streams where the current would be too strong for many other animals.   Their eggs are laid in batches inside a protective covering called a capsule or cocoon. The capsule may be attached to some solid object in the water, it may be laid in damp earth near water, or it may be carried about on the underside of the parent, according to the habits of the species.   All leeches are both male and female in the same animal (hermaphrodite) so that every individual lays eggs when it is mature.   Most kinds seem to live very well in aquaria ; they need very little attention since they do not feed often.   Care should be taken in the case of the horse leech, the *Herpobdellas*, and *Trocheta subviridis* to cover the top of the aquarium with muslin, as all these species leave the water fairly readily.

The following pages give a short description of each of the British fresh-water leeches with notes on any interesting features of their natural history.   If you compare the head of any particular leech which you find with diagrams on page 60 and its other features with the notes, then you should have little difficulty in identifying the species if you so wish.

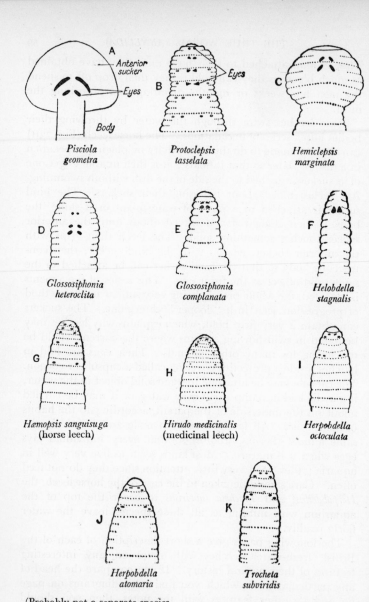

Fig. 24.—Heads of British fresh-water leeches. B-K modified after Harding.

A Anterior sucker
Eyes
Body

*Pisciola
geometra*

B Eyes

*Protoclepsis
tasselata*

C

*Hemiclepsis
marginata*

D

*Glossosiphonia
heteroclita*

E

*Glossosiphonia
complanata*

F

*Helobdella
stagnalis*

G

Hæmopsis sanguisuga
(horse leech)

H

Hirudo medicinalis
(medicinal leech)

I

*Herpobdella
octoculata*

J

*Herpobdella
atomaria*
(Probably not a separate species
but a variety of H. octoculata)

K

*Trocheta
subviridis*

Family Rhynchobdellidæ

Leeches with a proboscis, no jaws, colourless blood.

SPECIES.	*Protoclepsis tesselata* (Fig. 23A, B, C, D,), 24B.
COLOUR.	Olive-green or brownish, transparent.
MARKINGS.	Back shows six longitudinal rows of yellow spots in large specimens.
EYES.	Eight, arranged in four pairs (Fig. 24B).
SHAPE.	Varies according to degree of contraction or expansion (see Fig. 23). Anterior sucker not visible from above. Hind sucker conspicuous, as broad as the body when the latter is extended. A very active leech constantly changing shape when moving.
SIZE.	Length up to 6 cm. when extended.
FOOD.	Said to feed on ducks and other water-fowl, invading the nasal cavities and the throat region, sometimes with fatal results to the birds.
BREEDING HABITS.	Egg capsules are laid attached to underside of the parent. Young may be as many as two hundred, they are also carried about by the parent.
OCCURRENCE.	Considered rare in Britain. Recorded from Shropshire. I have taken specimens in Surrey, the Lake district, and Renfrewshire. Presumably this species is distributed by its hosts which may result in a local or a discontinuous distribution.
SPECIES.	*Pisciola geometra* (Fig. 24A).
COLOUR.	Greenish or brownish-red.
MARKINGS.	Eight longitudinal rows of whitish spots round the body.
EYES.	Two pairs at base of head sucker.
SHAPE.	Long and narrow. More cylindrical than any other leech. Both suckers very prominent with a frilled appearance due to radiating markings. Never contracts itself very much. A very active species which loops its body very obviously as it progresses.

SIZE.   Length 2-5 cm.

FOOD.   Attacks fish, such as trout and the Miller's thumb (*Cottus gobio*).

BREEDING HABITS.   Eggs are laid inside dark brown elliptical cocoons which are attached to stones or water-weed. Length about 1·5 cm.

OCCURRENCE.   Seems to be not uncommon in streams. I have taken many specimens in Hertfordshire, where they were found on the undersides of stones surrounding the nests of the Miller's thumb. This ectoparasite fills much the same rôle as do bed-bugs in modern civilisation, living in the "homes" of its hosts.

SPECIES.   *Glossosiphonia complanata* (Fig. 23F, G, 24E).

COLOUR.   Dark greenish-brown.

MARKINGS.   One pair of very distinct longitudinal rows of cream and brown-coloured spots on either side of the mid-line on the back. One pair of lateral rows of cream spots, making six rows in all. The spots are subject to variation.

EYES.   Three pairs, first pair small and sometimes absent.

SHAPE.   Elongate oval with narrow head end (Fig. 23F, G). Head sucker not visible from above. Sluggish in its habits, curling up into a ball when treated roughly.

SIZE.   Length about 3 cm. when moderately extended.

FOOD.   Fresh-water snails and "blood-worms" (*Chironomus* larvæ).

BREEDING HABITS.   Eggs laid attached to some solid object and enclosed in a transparent colourless cocoon.

OCCURRENCE.   Very common in Britain. Found in stagnant pools and running water.

SPECIES.   *Glossosiphonia heteroclita* (Fig. 24D).

COLOUR.   Clear amber-yellow.

MARKINGS.   Back often covered with small brownish or blackish spots.

EYES.	Three pairs. First pair may be unpigmented.
SHAPE.	Elongate oval with narrow head end. Head sucker not visible from above.
SIZE.	Length at rest 1-1·5 cm.
FOOD.	Water-snails.
BREEDING HABITS.	Eggs, as many as sixty, attached to underside of parent. Laid in June and July.
OCCURRENCE.	Common in England in slow-running or stagnant water. Rare in Ireland. Not recorded from Scotland.
SPECIES.	*Helobdella stagnalis* (Fig. 24F).
COLOUR.	Transparent greenish-grey or pinkish.
MARKINGS.	No definite arrangement of spots. Back surface finely speckled with black.
EYES.	One pair placed close together (Fig. 24F).
SHAPE.	Elongate oval, very narrow at head end. Head sucker not obvious from above.
SIZE.	A small leech, length about 1 cm.
FOOD.	Fresh-water snails, "blood-worms," small annelid worms.
BREEDING HABITS.	Eggs carried about on under-surface of parent.
OCCURRENCE.	Very common in ponds and slow-moving streams all over Britain.
SPECIES.	*Hemiclepsis marginata* (Fig. 24C).
COLOUR.	Ground colour, pale yellow, back surface variegated with orange, lemon-yellow, reddish-brown and bright green. Usually more green than any of the other colours. A yellow form also exists.
MARKINGS.	Green or reddish-brown transverse stripes. Usually seven series of yellow spots down the back. No spots on the under-surface.
EYES.	Two pairs.
SHAPE.	Anterior end seems to have a definite " head " due to shape of front sucker.
SIZE.	Length 1·6-1·8 cm. at rest.
FOOD.	Feeds on fish ; said to attack small kinds of carp.

| BREEDING HABITS. | Eggs are attached to under-surface of parent. |
| OCCURRENCE. | Rare in England. Recorded from Histon, Cambs., and the Shropshire Union canal. |

Family Gnathobdellidæ

Fresh water and land leeches with red blood, without an extensible proboscis, usually with jaws. Head sucker not distinct from the body below.

SPECIES.	The horse leech (*Hæmopsis sanguisuga*), (Fig. 23H, 24G).
COLOUR.	Varies from blackish-green, through olive-green and yellow-green, to brown.
MARKINGS.	Upper and lower surfaces densely spotted with black.
EYES.	Five pairs of eyes.
SHAPE.	A long narrow soft-bodied leech, head end very much attenuated. Central part of body has nearly parallel sides. Much less contractile than the medical leech (*Hirudo medicinalis*) with which it is easily confused. Jaws have between eleven and eighteen pairs of blunt teeth, but these cannot pierce the human skin.
SIZE.	Average length at rest 2·5-3·5 cm. Specimens up to 15 cm. when extended are quite common.
FOOD.	Carnivorous. Eats earthworms, snails, insect larvæ, tadpoles, and even its own species.
BREEDING HABITS.	It leaves the water to lay its cocoons in damp earth.
OCCURRENCE.	Common all over the British Isles. Found chiefly in mud at the bottom of ponds, or on stones in streams.

| SPECIES. | The Medical leech (*Hirudo medicinalis*), (Fig. 24H). |
| COLOUR. | Usually olive-green above variegated with red, yellow, orange, and black. |

MARKINGS.	Pattern on back is variable. Generally three pairs of longitudinal stripes of a reddish-brown or yellow colour ; these stripes are interrupted by black spots.
EYES.	Five pairs.
SHAPE.	Similar to that of the horse leech except that it can contract itself much more.
SIZE.	Length at rest about 3-3·5 cm.
FOOD.	Jaws have a large number of sharp teeth, these can easily pierce the human skin.
BREEDING HABITS.	The cocoons are laid in damp earth.
OCCURRENCE.	Used to be common in England, and was extensively collected for use by medical men. Later, supplies of this leech came entirely from abroad and the species was considered extinct in this country (1910). More recently it has been discovered in large numbers in the New Forest district.
SPECIES.	*Herpobdella (Nephelis) octoculata*, (Fig. 24 1).
COLOUR.	Dark brown, sometimes yellowish or reddish-brown.
MARKINGS.	Blackish markings.
EYES.	Four pairs.
SHAPE.	Body very elongated, much narrower than either the horse leech or the medical leech.
SIZE.	Length at rest about 3-4 cm.
FOOD.	Small worms such as Tubifex (p. 53) and Planarians (p. 24).
BREEDING HABITS.	Eggs are laid in dark brown transparent cocoons like those of *Herpobdella atomaria*. They are attached by both ends to solid objects. They are readily laid in aquaria.
OCCURRENCE.	Very common in running or stagnant water all over Britain.
SPECIES.	*Herpobdella atomaria* (Fig. 24J).
COLOUR.	Usually greenish-brown, paler underneath.

5

MARKINGS.   Reddish or yellowish spots on every ring except those near the head end. The surface is covered with a black reticulate pattern.

EYES.   Four pairs, as in *H. octoculata.*

SHAPE.   Very closely resembles *H. octoculata.* Considered by some to be a variety of this species only and not a distinct species.

FOOD.   Same as *H. octoculata.*

BREEDING HABITS.   Egg cocoons are brown and transparent (Fig. 23E). They are attached to solid objects in the water.

OCCURRENCE.   Very common. Found along with *H. octoculata* in running or stagnant water.

SPECIES.   *Trocheta subviridis* (Fig. 24K).

COLOUR.   Greyish-green or reddish, paler underneath.

MARKINGS.   Usually two brown lines along the back.

EYES.   Four pairs.

SHAPE.   Very long and worm-like, nearly cylindrical in cross-section.

SIZE.   Length at rest 8-10 cm.

FOOD.   Carnivorous, living on earthworms and insect larvæ. It is amphibious, leaving the water to crawl on land in search of food.

BREEDING HABITS.   Lays flat brown elliptical egg capsules like those of *Herpobdella*, attached to foreign bodies in the water.

OCCURRENCE.   No record from Scotland or Ireland. It has been found in the Zoo gardens, London, in Surrey, Herts, Hampshire, Sussex, and the Manchester sewage works.

Books for further reference—

Friend, Hilderic. The Story of British Annelids. London, 1924. Michaelson (1909). Die Süsswasserfauna Deutschlands. Heft 13. Brauer. A. Die Süsswasserfauna Deutschlands. Heft 13. Jena.

Harding, W. A. The British Leeches : Parasitology. 1910.

Boisen, Bennike, S. A. (1943). Contributions to the Ecology and Biology of the Danish Freshwater Leeches. (*Hirudinea*).

Folia Limnologica Scandinavica. No. 2.

CHAPTER 7

THE ARTHROPODA : CRUSTACEA

The Arthropods in their numbers and in their variety of species far exceed all the rest of the animal kingdom put together. Of the animals which live in fresh-water a high proportion belong to this phylum. The name " Arthropod " means jointed limbs, the most characteristic feature of the phylum being the fact that the numerous limbs on the body are made up of a number of separate segments. In many Arthropods the body shows some external indication of segmentation like the body of a worm, and to all or some of the segments are attached a pair of limbs. The limbs near the front end of the body have usually a sensory function, as in the antennæ or " feelers " of butterflies. The limbs behind the antennæ are usually modified into structures to seize and grind up food ; these are collectively called the jaws or mouth-parts, of which there may be from three to six pairs. Behind the mouth-parts come the limbs used for locomotion (i.e. walking or swimming). The skin on the surface of the animals contains a particular substance called chitin and is waterproof. Frequently most of the skin tends to be rather hard and brittle, and movement would be difficult were it not that between the segments on the body and the leg joints there are thinner, more flexible areas not impregnated with hard substances. The skin may be made harder still by the inclusion in it of lime salts, which are found for instance in the lobster and crayfish. The hard external covering of the Arthropods forms a supporting skeleton outside the body and gives attachment for muscles ; there is very little internal skeleton. The whole arrangement is the reverse of a vertebrate animal where the skeleton is inside with the soft tissues on the outside. The chief difficulty about an external skeleton is that it limits growth if it is not extensible. You therefore find that Arthropod animals all moult their chitinous skins periodically to allow for growth. In the arrangement of the internal organs the nervous system is much like that of an earthworm, but the blood does not circulate in a closed series of blood-vessels as in the Annelida. There is usually a tubular heart close to the surface on the back of the animal, which pumps blood forward into the head region, from here the blood passes through a series of large spaces which surround the chief organs. Eventually it enters the back and sides of the heart and is pumped forward again.

The fresh-water Arthropods fall into three main groups which will be treated separately. They are the Crustacea, the Insects, and the Arachnids.

The Crustacea are a large group of " jointed limbed " animals most of which live in water. A large number of different kinds are found in the sea (crabs, prawns, shrimps, etc.), but the fresh-water forms, though fairly numerous, have fewer species than those living in the sea. In addition to the general Arthropod characters of segmented bodies, numerous paired jointed limbs, and a chitinous skin, crustacean animals have two pairs of antennæ in front of the mouth followed by small limbs modified for feeding behind and around the mouth, and the body is usually divided into three regions, the head, thorax, and abdomen, of which the head may often be fused with part of the thorax forming the " cephalothorax." Since they nearly all live in water they breathe by means of gills, or if they are small through the general surface of their bodies.

The head is made up of six segments fused together, but the individual segments can only be seen in the embryo. This part of the body carries eyes which are often of the compound type and are sometimes placed on the end of movable stalks. The paired jointed limbs may be fairly numerous, sixteen to twenty pairs or even more. They are attached to the head (antennæ and feeding limbs), the thorax (where it is usually easiest to see that each segment has one pair of limbs), and sometimes the abdomen. The limbs may be used for a number of functions, such as feeding, walking, swimming, breathing, or as a sensory organ. As a rule the limbs are not all alike in shape, and this difference is connected with their various functions. A number of the Crustacea feed on small particles which they collect from the surrounding water by moving their limbs to and fro. Others use some of their limbs to seize and tear up their prey.

The skin (cuticle) on the surface of the body is generally hard except at the joints, this hardness being partly due to lime salts which are deposited in it. Many species have numerous bristles (setæ) arising from the surface of the skin, more particularly on the limbs. The whole of the skin, complete with bristles, is shed at intervals to allow for growth in young animals, and it continues to be moulted, though less often, when the animals are adult. The joints between the segments on the body and limbs are easy to see if the animals have thick hard skins like the crayfish, but the divisions are not easily seen when the cuticle is thin, and in the Water-fleas and Ostracods the body is enclosed in a kind of bivalve shell (or *carapace*) so that it is impossible to make out the segments at all.

Smaller kinds of Crustacea have fairly transparent bodies so that in many of them the internal organs are visible. It is usually possible to see the alimentary canal with its glands and to watch the heart beating in living specimens. The blood corpuscles in the blood may even be watched circulating through the body in a series of spaces which surround the principal organs (the hæmocœle), there being no closed system of blood-vessels in Crustacea.

Nearly all species are represented by males and females though a few parasitic types are hermaphrodite. Often males are found only at certain times of the year. In these cases the eggs laid by the females develop without fertilisation during the season when there are no males. Nearly all eggs, whether requiring fertilisation or not, are either laid into a brood pouch or else are attached to the surface of the parent where they continue to develop. Some eggs contain a large amount of reserve food in the form of yolk which is used up as the young grow. The larger the amount of yolk the longer the young remain inside the eggs, and the more advanced (i.e. the more like their parents) they are when they hatch. Many of the fresh-water forms emerge from their eggs looking very like adults except in size ; this is particularly true of those living in running water. Those which inhabit ponds and lakes hatch from their eggs as a larva (young stage),

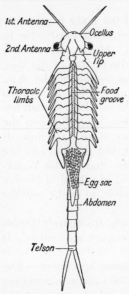

FIG. 25.—The Fairy Shrimp (*Chirocephalus diaphanus*), female, in swimming position, as seen from above.

called a *Nauplius* (see Fig. 99), which is not very like the adults. In every case growth and development take place with a series of moults, the young becoming more like the adult at each moult, and increasing in size until it is exactly like its parents and is capable of reproducing the species.

The Fairy Shrimps (Phyllopoda).— In Britain there is only one species of fairy shrimp which is found commonly, and this is *Chirocephalus diaphanus* ; it is quite large, semi-transparent and is a very fascinating animal to watch. Two more species have been reported from one or two localities occasionally, but they are certainly very rare.

Chirocephalus belongs to a group of crustacean animals which are considered to show more " primitive " features than any other Crustacea. This is because they have a very large number of segments to the body and most of the limbs are similar to one another in shape. This particular fairy shrimp is sometimes found in great numbers in small pools of water of the type that dry up in summer. Its chief features are shown in Figs. 25 and 26. The length of *Chirocephalus* is about 2·5 cm., and when swimming it always lies on its back. The animal is constantly in motion when alive, and the same graceful undulations of its leaf-like limbs which cause it to swim also produce a current of water which brings food par-

ticles to the mouth and oxygen to the tissues. The head is
clearly marked off from the rest of the body ; it bears a pair
of large black eyes on stalks. The first pair of antennæ are
very thin and unjointed, the second pair broad and fat in the
female, while in the male they are very large, complicated
structures (see Fig. 26). The three pairs of limbs acting as
jaws are closely placed round the mouth and are not easily seen
with the naked eye as they are covered by an upper lip. Behind
the "jaws" come eleven pairs of leaf-like limbs followed in the
female by a single ventral egg pouch. The abdominal part of
the animal is straight and narrow ; it begins with the segment
carrying the egg pouch, and there are six more segments ;
at the end is a small tail which forks in two. The whole
animal is transparent, bluish-green or almost colourless

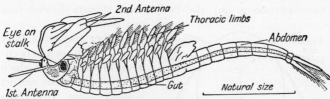

Fig. 26.—A male Fairy Shrimp (*Chirocephalus diaphanus*) in swimming
position. Adapted from Baird.

except that the tips of the limbs and the tail have a reddish
tinge ; the eyes are black, and the gut may be coloured green
by its partly digested contents. The food particles are strained
out of the water by the limbs and are carried towards the
mouth along a groove between the bases of the feet. They
consist largely of one-celled green plants, one-celled animals,
and organic débris. The gut, which is plainly visible through
the body wall, is a straight tube leading from the mouth to
the anus. In the head region there is an organ which com-
municates with the gut which is called the liver ; it probably
helps to digest and absorb food.

The eggs, which are laid into the brood pouch of the female,
have strong shells, and when the female dies these eggs can
survive for a long time in dry mud (see Fig. 27B). This explains why *Chirocephalus* can reappear in large numbers in

a pond which has been dried up the previous summer. When the eggs hatch the larva which comes out is a Nauplius (see

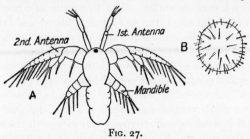

FIG. 27.

A. Newly hatched Nauplius larva of the Fairy Shrimp (*Chirocephalus diaphanus*. After Baird). B. Egg of Fairy Shrimp.

Fig. 27A), but it is an older stage than the nauplii of Copepod Crustacea (see p. 99), the body being already divided into two parts.

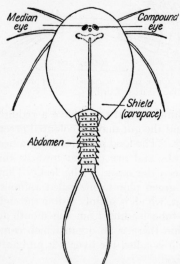

FIG. 28.—*Apus cancriformis*, dorsal view. After Borradaile.

Another species of fairy shrimp which used to occur in the British Isles but which appears now to be extinct is *Artemia salina*. In appearance it is like *Chirocephalus*, but the abdomen has only six segments. *Artemia* does not live in fresh water but in salt lakes and salt marshes where the amount of salt is much greater than it is in sea-water. It is still common on the Continent.

A third species which is sometimes found in this country is *Apus cancriformis*. This animal has a totally different general appearance from *Chirocephalus* because more than half the body is covered by a shield-shaped structure (the carapace) (see Fig. 28). The eyes in Apus are not placed on the end of stalks and both pairs of antennæ are much reduced. The limbs are all alike, leaf-shaped, and very numerous. It lives in the same kind of fresh-water pools as does *Chirocephalus* and it is said to eat insect larvæ. Males are very uncommon, and it is assumed that the eggs usually develop without being fertilised.

The Water-fleas (Cladocera).—The water-fleas are a large group of crustacean animals of which many kinds are found in fresh water. Most of them move through the water with a series of hops or jumps, and this is presumably why they are called " fleas." The size of these animals is, with one or two exceptions, anything from 1 mm. long up to about 3 mm., but what they lack in size may be made up in numbers, for they may be so abundant in a shallow pool that the water is literally hopping with them.

The general shape of a water-flea is shown in Figs. 29, 30. The head is slightly separated off from the main part of the body and possesses a single median conspicuous compound eye, which can be seen to rotate if the animal is observed alive under a lens. A small simple median eye (an ocellus) may also be present. The first antennæ are usually small and serve chiefly as sense (smelling) organs. The second antennæ are generally very large ; they are the organs of locomotion, their movements causing the animals to hop through the water. The chief characteristic of the water-fleas is that the main part of the body is enclosed in a kind of shell (carapace) made of the animal's skin ; this opens down the front like a jacket, giving the appearance of a shell with two valves, but it is really made of one piece. The whole of the animal's body behind the head, including the limbs, is enclosed by this structure in most forms. Some kinds have part of the abdomen projecting beyond the shell, and there are four British species which have the shell so much reduced that only the brood pouch on the back is covered, the rest of the body and the limbs being free. The limbs forming the jaws are small and are not easy to see clearly ; those limbs which follow behind the jaws are much larger, they can be seen waving to and fro inside the shell, because the latter is transparent. In typical water-fleas all or most of the five or six pairs of limbs are leaf-like in shape; their movement causes a current of water to pass between the limbs and out between the shell valves. This current helps the animals to breathe (which they do mainly through the surface of their

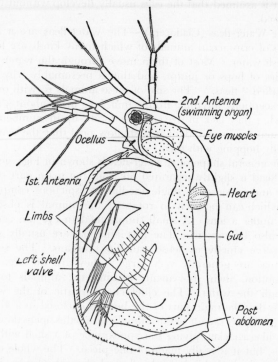

FIG. 29.—A common Water-Flea (*Simocephalus*).

limbs and through the surface of the inner side of the shell), and it also carries with it small particles of food, such as one-celled plants which are filtered off by the legs and passed forward to the mouth, edible particles being ground up by the jaws and pushed into the mouth. The posterior part of the body has no limbs; it is often edged with spines and usually ends in large curved claws.

Various internal structures are easily seen through the transparent covering of the body. The most obvious is the gut which is often coloured green for a large part of its course by the plants on which the animal is feeding. The gut may be a straight tube leading from the mouth to the anus, or it may have one or two coils near the middle. In the head region a pair of blind sacs or cæca are often present, they lead out from the gut and may also be coloured green ; they are thought to help in digesting the food. The heart, which is a very thin transparent sac, is placed dorsal to the stomach.

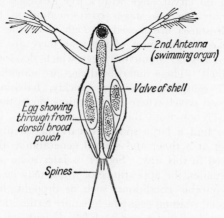

FIG. 30.—A Water-Flea, ventral view.

It contracts and expands during life at the rate of about 150 beats per minute pumping the blood forwards into the head region. The action of the heart results in a circulation of the blood round the body, and if you watch the heart beating under a microscope you will probably be able to see the blood-cells (corpuscles) and follow their course round part of the body. Other conspicuous features which show through by transparency are the powerful muscles which move the second pair of antennæ (the swimming organs) and the eggs or embryos in the brood pouch of some adult females.

The large eye of the water-flea probably does not perceive

a clear picture of the objects round the animal, at most it will only appreciate shadows of objects ; but it results in the flea being sensitive to light. These animals prefer light of a definite strength, not too strong or too weak ; on a dull day they will be found nearer the surface of the water than on a sunny day, while young specimens seem to prefer a stronger light than older ones.

All water-fleas lay their eggs into a brood pouch on the dorsal side of the female ; inside this the eggs develop, and when the young hatch they are miniatures of the adults, there being no larval stage with a free existence (except in *Leptodora*), because the stages corresponding to the larvæ of other Crustacea are passed through while the animal is still in the egg. For most of the year there are no males, and the females lay numerous eggs which develop without being fertilised. These unfertilised eggs are sometimes called "summer" eggs, they develop quickly, hatching in the female's brood pouch where for some time longer they may be fed by a secretion from the wall of the pouch. It is quite common to find a large specimen carrying twenty or even more young at a time, and several generations, all females, are produced in this way. Sooner or later some of the eggs hatch into males, the appearance of which may be associated with unfavourable conditions, such as drought, hunger, or cold. Special resistant eggs which require fertilisation are then produced by the females, and each female with few exceptions produces only one or two of these at a time. Such eggs remain protected by part or all of the female's skin when she next moults, and as a rule the brood pouch develops thickened walls which are dark in colour (see Fig. 35). This part breaks free from the rest of the skin when moulting occurs and it encloses the eggs which lie dormant for some months at the bottom of the pond. The thickened area round the brood pouch is called the "ephippium," because it is often saddle-shaped ; the eggs inside eventually hatch into females which then start the cycle over again. Males are seldom common, in some species they have never been seen ; those which are known look very like the females of the same species but they are smaller with larger first antennæ.

The length of life of a water-flea may be anything from a few weeks to six months. They live mainly in shallow weedy pools, in the backwaters of lakes, or among débris in almost any type of still fresh-water. They are eaten by a very large number of other animals ; small specimens are eaten by *Stentor* (a ciliate protozoan), *Hydra*, and flatworms to mention only a few ; larger specimens are eaten by newts, the young of many fish, and in captivity they serve as a good diet for other Crustacea such as young crayfish. Since water-fleas exist in very large numbers they are important in the food relations of fresh-water animals because they make the food obtained from the minute plants on which they feed available to larger carnivorous creatures.

There are about eighty different species of water-fleas in Britain but many of them are rather rare. They belong to seven families, and the particular characteristics of each family are given below along with drawings of some of the more common species.

Family Sididæ

Body and feet covered by bivalve shell (carapace). Six pairs of feet all leaf-like except the last pair. The swimming antennæ divided into two jointed portions with the joints flattened and with many lateral and terminal hairs.

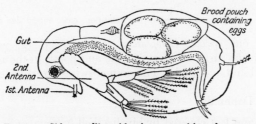

FIG. 31.—*Sida crystallina*, side view, actual length 3 mm.

There are five British species belonging to this family of which *Sida crystallina* is probably the commonest (see Fig. 31). It is found among weeds in ponds all over Britain, but it is not usually very abundant.

Family Holopedidæ

Hind part of the body and the feet not completely covered by the shell (carapace). The whole animal much flattened from side to side and enclosed in a clear gelatinous case which is open below. Six pairs of feet, all leaf-like except the last pair. Swimming antennæ with only one branch of three joints and ending in three long feathery hairs.

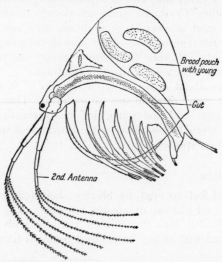

FIG. 32.—*Holopedium*, removed from gelatinous case, actual size 1·5 mm.

Only one species *Holopedium gibberum* (Fig. 32) which is found in the surface waters of lakes and ponds in the north of England and Scotland. The animal swims on its back.

Family Daphnidæ

Body and feet covered by bivalve shell (carapace). Five or six pairs of feet of which the first two pairs are *not* leaf-like. Second antennæ usually very small. Post abdomen generally flattened with anal spines. Gut has a pair of hepatic cæca.

Twenty-four British species (and many varieties) of which the commonest genera are *Daphnia* and *Simocephalus*. *Daphnia*

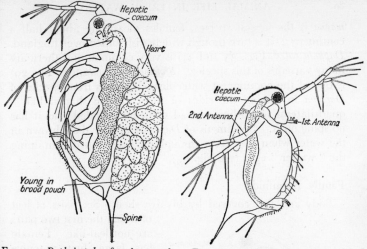

FIG. 33.—*Daphnia pulex*, female, actual length 2 mm.

FIG. 34.—*Daphnia pulex*, male, actual length 1 mm.

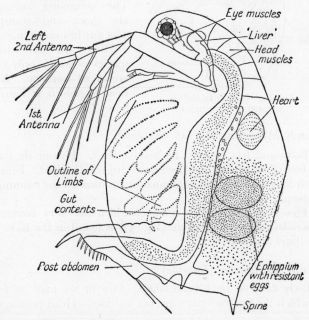

FIG. 35.—*Daphnia*, side view, actual length, 2 mm.

magna is the largest species, females measuring about half a centimetre ; it is rare though widely distributed over England. *Daphnia pulex* (Figs. 33 and 34) is very common and there are many varieties of this species. *Simocephalus vetulus* (see Fig. 29) is one of the commonest water-fleas, it differs from species of *Daphnia* in having the sides of the carapace shaped like a quadrilateral in outline instead of being rounded and it has no spine on it. Specimens of *Daphnidæ* appear dark brown in the water when they have the ephippia developed containing the " winter " eggs (see Fig. 35).

Family Bosminidæ

Body and feet covered by bivalve shell. Six pairs of feet of which the first two pairs are not leaf-like. Female has large first antennæ which are immovable. The swimming antennæ are short and stumpy. The gut has no cæca.

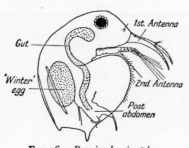

FIG. 36.—*Bosmina longirostris.*
Adapted from Scourfield.

There are six British species of which only two are at all common. *Bosmina longirostris* is figured here (Fig. 36).

Family Macrothricidæ (*Lyncodaphnidæ*).

Body and feet covered by bivalve shell. Five or six pairs of feet of which the first two pairs are not leaf-like. Female with long movable first antennæ in addition to the swimming antennæ.

Eleven British species none of which are very common. *Macrothrix rosea* figured here (Fig. 37) is found in the lakes of Scotland and Ireland.

Family Chydoridæ (*Lynceidæ*)

Body covered by bivalve shell. Five or six pairs of feet of which the first two pairs are not leaf-like. Head prolonged

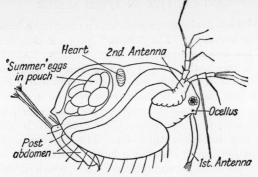

FIG. 37.—*Macrothrix rosea*, length 0·7 mm.   Adapted from Birge.

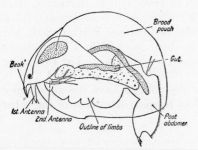

FIG. 38.—*Eurycercus lamellatus*, 3 mm.

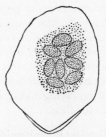

FIG. 39.—Ephippium of *Eurycercus lamellatus*.

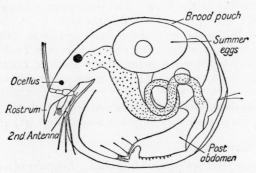

FIG. 40.—*Chydorus*, 0·3 mm. long.

in front into a beak-like process which covers the base of the swimming antennæ. The "beak" projects downwards in front of the first antennæ.

This is much the largest British family with forty-one species, but of these only two are very common. *Eurycercus lamellatus* (see Fig. 38) is easily recognised on account of the large size ; the female is about 3 mm. long which is much larger than any of the other Chydoridæ. It is found all over the British Isles in still water. The ephippium (Fig. 39) is peculiar because it contains a large number of " winter " eggs (i.e. ten to thirteen) instead of one which is the usual number for other Chydoridæ. *Chydorus sphæricus* is a small member of this family, but it is the commonest of all the water-fleas (Fig. 40). It is found in every type of still water, from small moorland pools and horse troughs to reservoirs and lakes. You may not come across it at first because it is so small, but as soon as you recognise it, then you will find it in almost every sample of still water which you examine. The swimming movements of this family (Chydoridæ) are carried out in such quick succession that the animals have the appearance of scurrying along instead of hopping like most of the other water-fleas.

Family Polyphemidæ

Body and feet not covered by a shell. Four pairs of feet with spines not leaf-like. The body is very short with an enormous eye in the head. The shell forms a brood pouch. There is a long spine at the end of the body.

Three British species found in lakes and pools. *Polyphemus pediculus* is found in the south, east, and north of England, and *Bythotrephes longimanus* is found in the north of England and Scotland and Ireland. The surface water of reservoirs sometimes contain large numbers of *Polyphemus* in July and August. The animals appear as enormous numbers of little black specks on account of their very large black eyes (Figs. 41, 42).

Family Leptodoridæ

Body and feet not covered by a shell. Six pairs of feet which are not leaf-like, the first two pairs being very long. The shell forms a brood pouch. The body is long and cylindrical and possesses very large swimming antennæ (Fig. 43).

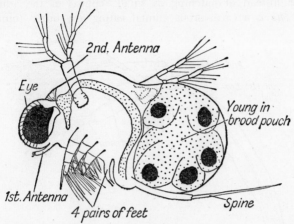

FIG. 41.—*Polyphemus* from Windermere, actual size, 1·4 mm.

Only one British species, *Leptodora kindtii*, which is found in lakes and large reservoirs all over the country except in the

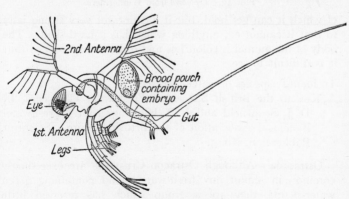

FIG. 42.—*Bythotrephes* from Windermere.

south of England. It is found in the surface waters, the only easy way to get specimens being to use a tow-net attached to

a small boat. This is the largest water-flea, the female is
nearly 2 cm. long. The eggs hatch into advanced Nauplius
larvæ instead of hatching as small replicas of the adults.
Leptodora is a carnivorous animal eating any other animals

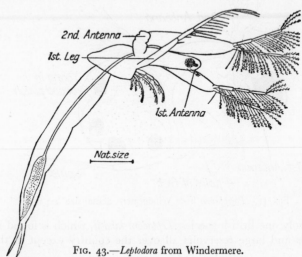

FIG. 43.—*Leptodora* from Windermere.

of which it can get hold, but it has not got very strong jaws
so that it cannot eat anything which has a hard shell. The
body of the animal is colourless and quite transparent so that
it is difficult to see.

To identify species of water-fleas see

A Key to the British Species of Freshwater Cladocera, by
    D. J. Scourfield and J. P. Harding, 1941.   Freshwater
    Biological Association of the British Empire.   Scientific
    Publication, No. 5.

Ostracoda.—Although Ostracod Crustacea are exceedingly
common in almost any fresh-water pond containing green
water-plants, they are a group which has received little
attention in this country. Individual Ostracods are very
difficult to identify and only a specialist can undertake to
do so. For this reason no attempt will be made to give an

account of the different genera. All Ostracods are very much alike in structure, so that it is easy to recognise an Ostracod as such, even if you do not know its particular name.

The crustacean characters of an Ostracod may not be obvious at once because the whole of the body is enclosed by a two-sided shell. When the animal is disturbed the valves of the shell close and are kept tightly shut by means of muscles so that the antennæ, limbs, and everything else are inside. The shell may be quite hard, and it is seldom sufficiently transparent for much to be seen through it. It is only when the animal begins to move again that the shell opens slightly to allow two pairs of antennæ and one pair of limbs to project through the

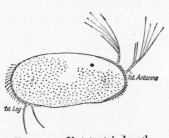

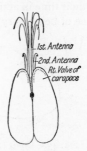

FIG. 44A.—*Herpetocypris*, length 2·5 mm.

FIG. 44B.—*Cypria* viewed from above.

opening. If you watch this happening under a lens then you will realise that the animal has some crustacean features.

Most Ostracods in Britain are very small, the largest is between 3 and 4 mm. long while the majority are about 1·5 mm. long. Viewed from the side they are bean-shaped (see Fig. 44A), while from above they look more like an egg (see Fig. 44B). The shell is often brown with blackish markings, or pale yellow with green markings, or sometimes almost peacock-blue. The colour is thought to match the surroundings of the animal so that specimens from a pond with much green vegetation will be greenish in colour, while those from a pond with brown debris at the bottom will be more likely to be a brown species. The black eyes are usually fused into one near the back side of the animal : they have a bronze iridescence when the light catches them in a certain way. Some forms are active swimmers and can move with considerable speed. The antennæ have long hairs which are generally very numerous in swimming forms, because the swimming is done with the antennæ which are held as in Fig. 45 and then moved backwards and forwards alternately. All kinds are able to " walk " along the stems of aquatic plants or on the bottom of the pond. When walking the first antennæ are waved about as in swimming but more slowly, while the second antennæ and the first pair of feet are used for touching solid objects. Fig. 45 shows an Ostracod " walking " as seen from the side. Ostracods seem to spend a large part of their time in active motion, and they always have the appearance of being very " busy " as they scuttle about.

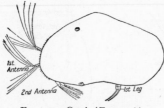

FIG. 45.—*Cypris* (*Eurycypris*), length 2 mm.

It is difficult to see any of the other structures of the body without killing the specimen and removing one valve of the shell under a dissecting microscope—a process which requires skill. In Fig. 46 the chief structures are shown as seen after removal of one shell valve. As well as the two pairs of antennæ there are three pairs of limbs used as jaws, and two pairs as legs. The body ends in a pointed structure called the *furca* which is sometimes pushed out through the shell opening. All species seem to moult very frequently, the cast skins floating up to the surface of the water. If one of these is examined the skin which covered the antennæ, legs, etc., will be found inside the shell valves. If you keep Ostracods in a small aquarium the surface of the water becomes covered with cast skins in quite a short time.

Ostracods eat anything, including decaying vegetation or small animals. In winter you may sometimes find that the red fruits of wild roses which have fallen into a pond have large holes made in them by Ostracods which will be found embedded in the tissues of the fruit. They are said to eat the

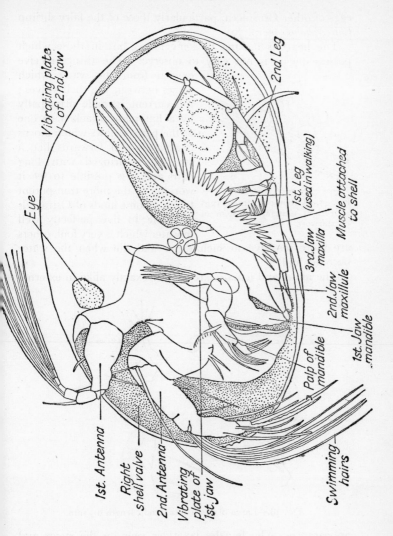

FIG. 46.—*Pionocypris vidua*, side view, left shell valve removed. Adapted after Cannon.

eggs of other Crustacea, particularly those of the fairy shrimp *Chirocephalus* (see p. 70).

The heart is lacking in some forms, but in those which possess one it is not easy to observe. The tissues receive

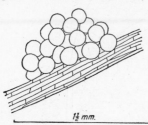

oxygen from the water which passes between the shell valves; this current is made principally by a flat plate attached to one of the pairs of jaws which moves backwards and forwards like a fan; it is fringed with long hairs and it is possible to see it working in the more transparent forms. Some kinds of Ostracods are able to live perfectly well in water which is very foul, others are more sensitive to such water and die when the water becomes foul.

FIG. 47.—Group of Ostracod eggs (bright orange), on stem of starwort.

The males and females are usually exactly alike in external

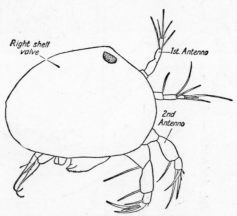

FIG. 48.—Larva of *Cypris* (Ostracod), length 0·3 mm.

appearance. The females lay their eggs on the stems and leaves of water-plants instead of carrying them about with them as most Crustaceans do. Shortly before a female is ready to lay her eggs these latter are often visible through the shell as large, round, white or orange-coloured bodies. If such a female is carefully watched she may be seen crawling about on water-plants, often stopping to inspect a particular place with her antennæ. When she has found a suitable place she deposits a single egg or a group of eggs stuck together. If you have a large number of female Ostracods laying orange-coloured eggs

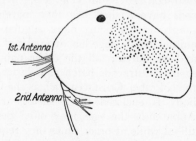

FIG. 49.—Young Ostracod, length 0·5 mm.

in a small aquarium, then the groups of eggs to the naked eye appear like patches of rust on the water-plants. If these eggs are kept in water under laboratory conditions of temperature they hatch in about six weeks to two months, but

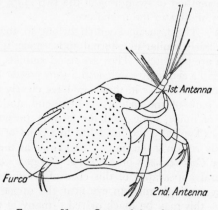

FIG. 50.—Young Ostracod, length 0·9 mm.

they are also capable of surviving in dry mud for twenty years or more, at the end of which time they will develop normally if the mud is placed in water. A very common green-and-yellow-coloured Ostracod (a species of Cypris,

Fig. 45) found in small ponds lays large numbers of bright orange eggs (see Fig. 47) during the months of March, April, and May. From the eggs reddish-coloured young Ostracod hatch out (Fig. 48). When newly hatched, Ostracods have a shell like their parents, but it is a different shape, usually being more angular (see Figs. 49, 50). The animal at this stage has two pairs of antennæ, one pair of jaws, and the furca ; all the other pairs of limbs develop later, so that Ostracods hatch as a larva which is really very like that of other Crustacea such as the Copepods, except that the body is enclosed by a shell ; this larva is a Nauplius. Apparently the eggs may often develop without being fertilised while at other times they require fertilisation.

Ostracods are so very common that you will probably come across them accidentally among mud or weed that you have collected for other animals. Where they are exceptionally abundant in small pools they may be very easily collected with a fine net. At first sight they may be confused with some of the smaller types of water-flea (such as the *Chydoridæ*) because the method of swimming of the two kinds of animals appears very similar to the naked eye.*

Copepoda.—The Crustacea which belong to the Copepod class are all fairly similar in external appearance, if we except those which are parasites. They are small animals, few being more than half a centimetre in length ; they possess the typical crustacean characteristics in that they have clearly segmented bodies, two pairs of antennæ at the head end, and numerous other pairs of limbs, some used for feeding purposes or for swimming. (Typically the head has five pairs of limbs and the mid-body region seven pairs.) A glance at the drawings of *Cyclops* or *Canthocamptus* on page 101 will give you an idea of the general appearance of this class of crustacean. Most of them have a single median eye near the upper surface of their heads—this may be black or red ; many of them have very numerous stiff bristles (setæ) arising from the surface of their bodies, particularly from their limbs. Colouring such as green, orange, or blue may be present in their bodies making such individuals look very attractive under the low power of a microscope.

* To identify Genera and Species of Ostracods, see Die Tierwelt Deutschlands (1938). Crustacea 3. Ostracoda. von Walter Klie. Jena.

Many Copepods are very common in the still water of ponds and lakes. Often one species will occur in enormous numbers in a particular stretch of water. They may be collected most easily by dragging a net of fine mesh (if possible made of bolting silk) through the water, or by taking a supply of thread-like algæ (filamentous green water-plants) and water-weed from the pond and placing it in a container of water, when specimens which have been entangled in the threads of the plant can be captured. Copepods (with the exception of some *Harpacticids*) are seldom found among mud; they are generally to be found swimming in the water, among algæ, or among damp moss. Occasionally you may come across a small ditch or pond that contains so many of these creatures that the water is like soup. They are often associated with water-fleas (Daphnia, see p. 79).

Little is known about the feeding habits of Copepods. Some are said to eat diatoms and algæ (unicellular microscopic plants). Other groups filter small particles out of the surrounding water by means of a complex sieving mechanism; they are called filter-feeders. Probably they select all particles suspended in the water of a certain size, and do not distinguish between palatable and unpalatable. Cyclops (see p. 97) is able to feed on relatively large dead animals. Copepods are themselves eaten by fish. They act as intermediate hosts for a few parasitic worms, and these parasites reach their second host when a fish eats the infected Copepod (see p. 41).

When the eggs of Copepods are laid by the female they remain cemented together in a mass called an egg-sac which is attached to the female's body. It is characteristic of some genera to have one egg-sac, as in *Diaptomus* (Fig. 52) or *Canthocamptus* (see Fig. 60), while two are always present in *Cyclops* (Fig. 54). Some Copepods lay two types of eggs at different times of year in these sacs. One kind develops quickly and are ready to hatch in about ten days; the other kind, known as resting eggs, are larger, and they may take many weeks to develop fully. These latter are resistant to the unfavourable conditions of cold or drought. They may be carried long distances by the wind, or by their being picked up with mud on birds' feet. Resting eggs are responsible for the dispersal of the species. Any one species of Copepod is commonest at the time of year when it is producing quick developing eggs ; this may be the winter or the summer ; for instance *Diaptomus castor* is commonest in winter. In some forms including one *Cyclops* the adults may encyst, and hibernate during unfavourable conditions.

When the eggs hatch the animals emerging from them do not resemble their parents very closely. They are much smaller, broader in proportion, have only a few pairs of limbs, and possess no " tail end " to their body. They are the Nauplius larvæ (Fig. 61) : there are small differences between the Nauplii of different species just as there are differences in the adults. The Nauplii go through a series of moults, up to six in number ; at each moult the skin is shed complete with setæ, the animals increase in size, and they show more of the adult characters. After these moults they reach the first " copepodid " stage (Fig. 62) and after about five further moults they become exactly like their parents. The time taken for a newly hatched Nauplius to develop into an adult varies with the temperature of the water ; in some it is from twenty-two to twenty-six days. Young Nauplii may easily be obtained by keeping a female with an egg-sac in a small amount of water and examining her each day until the eggs hatch. If she has been kept in a covered watch-glass or some other small container, it will be quite easy to find the Nauplii under a microscope. They are of course very small, about a tenth of a millimetre long ; they may also be colourless, the only conspicuous part of them being their eye. Quick developing eggs take roughly between five and twelve days to hatch, and the first moult may take place in less than twenty-four hours after hatching, so that cast skins may also be found. Male Copepods are as a rule smaller than the females. Their first pair of antennæ, or at least one of the antennæ is curiously jointed so that they can catch hold of the female by curling one antenna round the end of her abdomen.

Copepods are divided into three sub-groups (see Fig. 51) as follows :—

CALANOIDA. All these have long antennæ (at least as long as the abdomen, and having up to twenty-five segments) which are used as balancers. Abdomen clearly marked off from thorax. There are five British

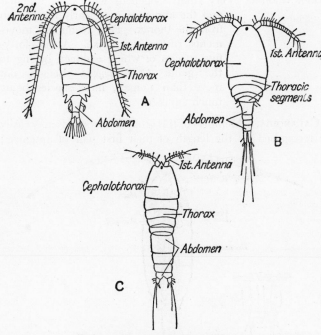

FIG. 51.—The three kinds of Copepod.

A. Calanoid Copepod (*Diaptomus*).
B. Cyclopoid Copepod (*Cyclops*).
C. Harpacticid Copepod (*Canthocamptus*).

genera with a total of twelve species. Length 1½-2½ mm. when adult.

CYCLOPOIDA. Antennæ much shorter than the Calanoida. Abdomen clearly marked off from thorax except in the parasitic forms, which may be unrecognisable as Copepods. About forty-six free-living species of which forty-four belong to the genus *Cyclops*. Length ½-3 mm. Two species are semi-parasitic and about seven are completely parasitic. Length of female parasites up to 7 mm.

HARPACTICOIDA. Very small Copepods with short antennæ and the abdomen not clearly marked off from the thorax. Poor swimmers, movements rather worm-like. About forty-five species of which eighteen belong to the genus *Canthocamptus*. Length not more than 1 mm., most species being much smaller.

CALANOIDA.—Members of this group of Copepods may easily be recognised by the length of their first pair of antennæ ;

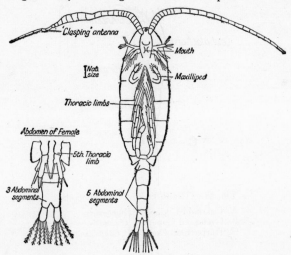

FIG. 52.—*Diaptomus*, male, ventral view, and abdomen of female.

these are equal to at leas. half the length of the body and are often longer (Figs. 52, 53). They are made up of from twenty-two to twenty-five segments. In the males one antenna is modified for grasping the female (Fig. 52). The hind part

of the body is distinctly divided off from the fore part. The
female carries a single egg-sac except in those few species which
lay their eggs into the water.

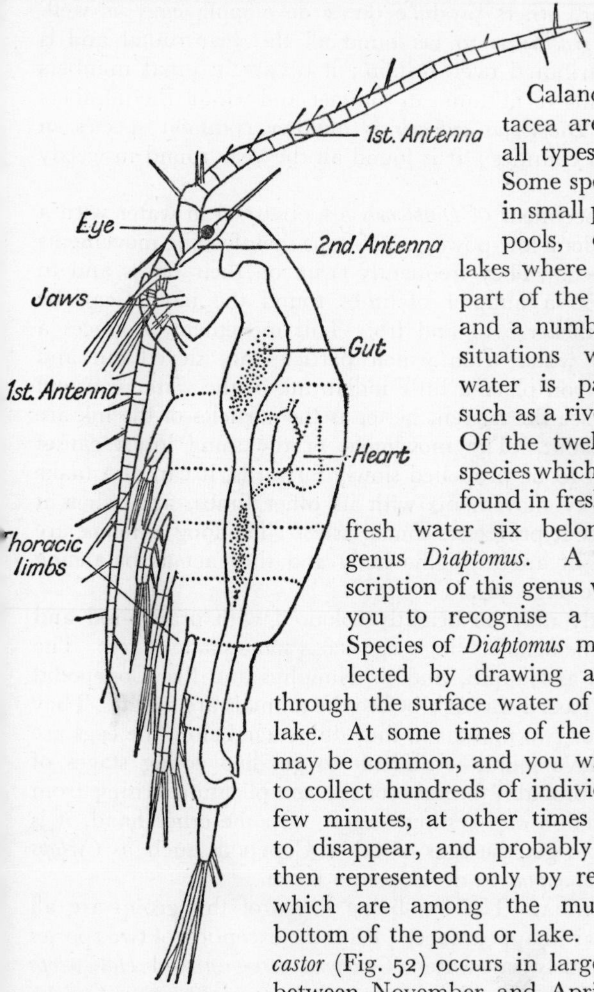

Calanoid Crustacea are found in all types of water. Some species occur in small permanent pools, others in lakes where they form part of the plankton, and a number inhabit situations where the water is partly salt, such as a river estuary. Of the twelve British species which have been found in fresh or nearly fresh water six belong to the genus *Diaptomus*. A brief description of this genus will enable you to recognise a Calanoid. Species of *Diaptomus* may be collected by drawing a fine net through the surface water of a pond or lake. At some times of the year they may be common, and you will be able to collect hundreds of individuals in a few minutes, at other times they seem to disappear, and probably they are then represented only by resting eggs which are among the mud at the bottom of the pond or lake. *Diaptomus castor* (Fig. 52) occurs in large numbers between November and April in small permanent ponds containing water-weeds. It is very rare or completely

FIG. 53.—*Diaptomus*, female, side view, length 1·5 mm.

absent during the summer months. Some species appear to produce only resting eggs so that there is only one generation in the year, others produce quick developing eggs as well. *Diaptomus gracilis* is to be found all the year round and is widely distributed over Britain ; it occurs in great numbers during spring or autumn, in summer and winter the numbers decrease. *Diaptomus vulgaris* is another common species in the eastern counties ; it is found all the year round in weedy pools.

If live specimens of *Diaptomus* are observed in water with a lens, or under a low-power microscope their feeding movements may be seen. They frequently swim on their backs and in this position a number of limbs round the mouth can be seen to " flicker " to and fro. This movement produces a current of water from which particles are sieved out and eaten. If you place a little indian ink in the water you will be able to see the current better as the particles of the ink are whisked about. The movement of the same limbs causes the animal to be propelled slowly forwards, it can also make sudden jerky movements with its other limbs, and when it makes these it progresses much faster. The long antennæ are held at right angles to the body and they act as balancing organs (see Fig. 52).

The body may be brightly coloured with orange-red and blue, and the female's single egg-sac is some shade of red. The eggs hatch as Nauplii, and six Nauplius and five Copepodid stages are passed through before the animals are adult. They may live for two months in the adult stage before the eggs are laid. I have found it difficult to get live young stages of *Diaptomus* probably because of the complication arising from the production of " resting " eggs. On the other hand, it is very easy to get the eggs of other Copepods such as *Cyclops* and *Canthocamptus* to develop into larvæ.

CYCLOPOIDA.—The free-living forms of this group are all very similar in appearance. With the exception of two species found in fairly salt-water (*Cyclopina norvegica* and *Halicyclops æquoreus*) all the free-living species belong to the genus *Cyclops*. Figs. 54 and 55 show two female *Cyclops* of different species. The body is much like that of the Calanoid *Diaptomus*, but

the anterior part is fatter and the antennæ are shorter.   The
limbs round the mouth are different for *Cyclops* is not a filter
feeder ; it seizes food particles, and it can apparently live
on the dead bodies of other animals, but very little is known
about its method of feeding.   *Cyclops* can swim slowly by means
of its two pairs of antennæ, or it can swim faster with jerks

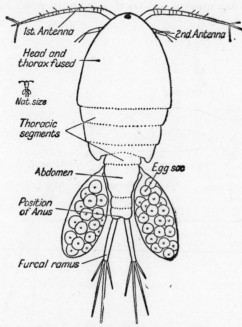

1st. Antenna

2nd. Antenna

Head and
thorax fused

Nat. size

Thoracic
segments

Abdomen

Egg sac

Position
of Anus

Furcal ramus

FIG. 54.—*Cyclops*, female, dorsal view.

by using other limbs on its thorax (see Fig. 55).   Members of
the genus *Cyclops* are very widely distributed and are to be
found in almost any type of still water all over Britain.   They
may also be very common in slow river and canals, particularly
among weeds.   The female carries two egg-sacs (see Fig. 54),
and she is readily recognised with the naked eye if egg-sacs
are present.   It is easy to obtain Nauplius larvæ from the
eggs if the female is isolated in a small dish for a number of

days until the eggs hatch. A dorsal view of a Nauplius larva of *Cyclops* is shown in Fig. 56. If you collect a large number of *Cyclops* and keep them in a small jar in the water in which they were found, and then you examine samples of the water carefully you will very likely find all stages in the life-history. As far as is known *Cyclops* do not produce resting eggs, all the eggs develop quickly so that if you find a pond where the

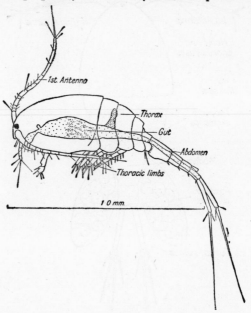

FIG. 55.—*Cyclops*, female, side view, length 1 mm.

females are mostly carrying egg-sacs, then you will be very likely to find young stages among your catch.

Parasitic Cyclopoida.—The two semi-parasites found in Britain live on the gill region of fish. *Ergasilus* (see Fig. 51) is apparently rare, and has been found on the gills of grey mullet. *Thersitina* (see Fig. 58) is found on the inside of the gill cover of sticklebacks which are living in slightly salt-water; it is sometimes very common, and forty specimens have been taken from one gill cover. In both these semi-

parasites all the young stages and the adult males are free-
living, it is only the adult females which are parasitic.   The
shape of the adult female is fairly like that of a typical Copepod,
the egg-sacs are however particularly large.

In the completely parasitic Copepods the males as well as
the females may be parasites, and the bodies of these animals
do not look at all like an or-
dinary Copepod.   Their eggs
hatch into Nauplii and the
beginning of their development
is just like that of any ordinary
Copepod, so that their curious

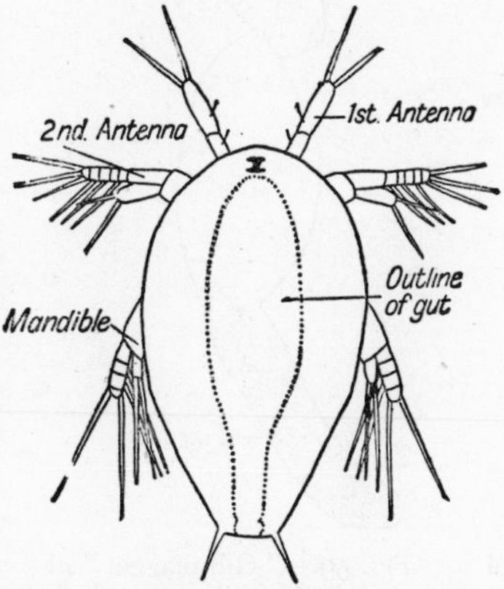

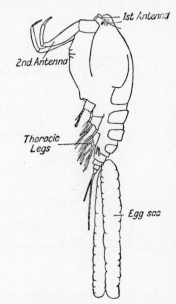

FIG. 56.—*Cyclops*, 1st stage Nauplius,
dorsal view.

FIG. 57.—*Ergasilus*, female, side view,
length 1 mm.   After Gurney.

adult shape must be considered a secondary development
due to their parasitic habit.   The " sea-louse " of salmon
(*Lepeophtheirus salmonis*) is found on salmon in many rivers
shortly after the fish have returned from the sea.   The female
is about 1·5 cm. long, the male only half a centimetre.   They
are most numerous round the anal region of the fish.   They
do not survive very long in fresh water because they are really
sea animals.

The " gill maggot " of salmon (*Salmincola salmonea*) (Fig. 59), is another parasitic Copepod. Only the female is known and breeding takes place in fresh water, but the parasite is not killed when the salmon migrates into the sea. Two other parasitic Copepods of the same genus have been found on the gills of grayling and trout in Yorkshire and the north of Scotland. They are both about 3 mm. long and similar to the "gill maggot" in shape. About three more species of

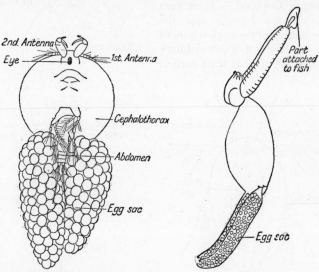

FIG. 58.—*Thersitina gasterostei*, ventral view of female, length 0·6 mm. Modified after Gurney.

FIG. 59.—" Gill maggot " of salmon (*Salmincola salmonea*), side view, length 7 mm. After Gurney.

parasitic Copepods may occur in Britain on fresh-water fishes, but they are certainly rare.

HARPACTICOIDA.—This group of Copepods contains animals of very small size, considerably less than 1 mm. being the average length for an adult. In some there is no division of the body into a thoracic and abdominal part, while in others it is not very distinct (see Fig. 60). The first pair of antennæ are quite short and are not used for swimming. In fact the method of locomotion is more like creeping than swimming ; the limbs are worked backwards and forwards, and the body bends and straightens with jerks. Their progress is slow, but among mud or fine strands of algæ they appear to move quite effectively. There is no doubt that these creatures may be tremendously common though they are often overlooked on account of their small size. They are to be found in ponds and lakes among water-plants or on the bottom amongst detritus. Some forms occur on damp moss, particularly Sphagnum moss, and others prefer slightly salt-water. I have found samples of a common alga *Vaucheria* which were simply alive with a Harpacticid belonging to the genus *Canthocamptus* (see Fig. 60).

The genus *Canthocamptus* will serve as an example of this group of Copepods. There are eighteen species in Britain and Fig. 60 shows a drawing of *Canthocamptus staphylinus*. This species is commonest in the winter months from about November to February, at which time many females will be found each with a single egg-sac. It appears to be very rare or absent in summer. I found that the eggs hatched in about eight days after laying, and the same female laid a further batch of eggs one week after the first had hatched. The larvæ when they hatch are of course very small, and it

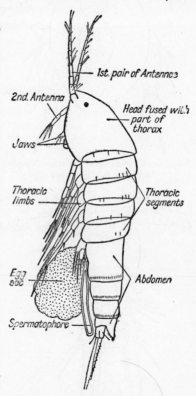

FIG. 60.—*Canthocamptus staphylinus*, female, side view, length 1 mm.

may take you some time to find them.   Females carrying egg-sacs have a curious structure attached to their bodies which is a spermatophore (see Fig. 60). This contains the sperm from a male and it remains attached to the body of the female until she dies.   The young stages are fairly like those of other Copepods (Figs. 61, 62) but they move rather differently.   The males have both their first antennæ modified for grasping hold of the female (Fig. 63). In other respects they are like the females.

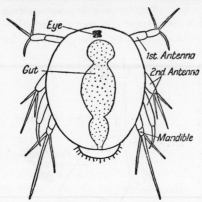

FIG. 61.—*Canthocamptus*, 1st stage Nauplius, larva just hatched from egg.

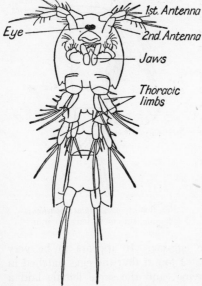

FIG. 62.—*Canthocamptus*, young stage, ventral view, 1st Copepodid stage. After Gurney.

To identify British Copepods see

British Fresh - water Copepoda, by R. Gurney. Vols. I, II, III, Ray Society, London, 1931.

The Fish Lice.— There is one family of Crustacea which are commonly known as fish lice (Argulidæ). They are sometimes grouped together with the Copepods because their structure is similar in many respects ; and sometimes they are placed in a group by themselves.   These creatures are all external parasites of fish and attach themselves most commonly to the gill region.   They are found on sea and fresh-water hosts, the species *Argulus foliaceus* being common on fresh-water fish in Britain (see Fig. 64).

Fish lice live by sucking the blood of their host.   Their bodies are disc-shaped, with a pair of saucer-shaped suckers near the head end on the ventral side, which are used to obtain

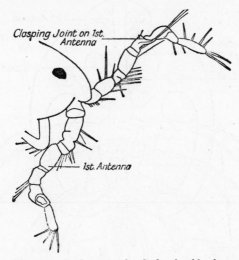

Clasping Joint on 1st. Antenna

1st. Antenna

FIG. 63.—*Canthocamptus*, head of male, side view.

a grip upon the surface of the host.   Because of their extreme flatness these animals offer little resistance to the water so that they are not brushed off by the current when the host makes sudden swimming movements.   Nearly all the parts of the animal are easily seen from either surface because the body is very transparent.   The males are slightly smaller and are more transparent than the females.   The latter often have a large opaque mass of eggs in the middle of the body. Both pairs of antennæ are reduced to small projections placed immediately in front of the eyes, which latter structures are very conspicuous. Between the ventral suckers is a poison spine, and just behind them is a suctorial proboscis for obtaining food from the host. Behind the mouth appendages there are five pairs of limbs ; the first pair are small but the others are large, and each limb has two feathery shafts at the end. These four pairs of limbs are flicked backwards and forwards spasmodically all the time the animal is attached to its host or any other object.

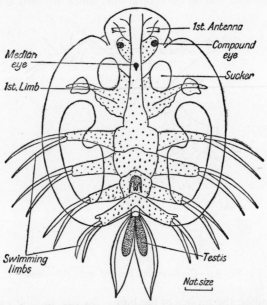

Fig. 64.—The Fish-Louse (*Argulus*), male viewed from above.

Fish lice are able to swim about freely. From time to time they attach themselves to fishes in order to feed ; one louse will attack several species of fish at different times. When swimming the feet are used as well as the expanses on the sides of the body. Their movement resembles very much that of a small flounder.

The females can certainly be fertilised by a male while not attached to the host, as this occurs in captivity. The eggs are laid on stones or other solid objects.   The British species *Argulus foliaceus* lays eggs in July and August, but it may well do so during other months in addition.

Isopod Crustacea.—The Isopoda is the scientific name for the group of animals commonly known as " Slaters." The wood-lice or slaters which are found under stones are among the few Crustacea which are really land animals ; there are closely related seashore forms and there are also two common

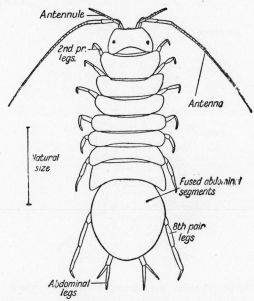

FIG. 65.—The Water-Louse (*Asellus*), adult specimen.

British fresh-water species.   These latter are called hog slaters or water-lice (*Asellus*).

A water-louse (*Asellus*) looks rather like an ordinary wood-louse, but it is less solidly made and the legs and antennæ are much longer (see Figs. 65 and 66).   The characteristic feature about the body is that it is flattened from above downwards ; it is so flat compared with its depth that when you are examining a water-louse by placing it on its back it is difficult to imagine how all the internal organs have room in which to work. These creatures are very common indeed in ponds or slow-moving streams, and they are sufficiently large to be seen easily in the water ; they do not swim, but crawl or climb about among weeds. Sometimes they are 2·5 cm. long but generally they are rather less. They live excellently in a small aquarium among water-weed, they breed readily, and they are not upset by the water becoming

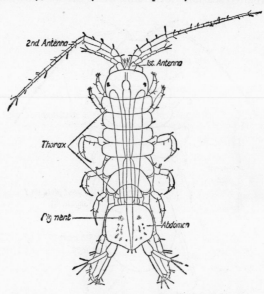

FIG. 66.—A young Water-Louse (*Asellus*), length 3 mm.

foul. They are in every way suitable animals to keep for demonstrating crustacean characters.

The colour of the body is a grey-brown due to pigment in the skin. The amount of pigment varies, sometimes the animals are nearly white, others may be quite dark, in which case it is impossible to see any internal structures. The under-side is a dirty white. If a live specimen is placed on its back in a small amount of water all the limbs can be seen ; they are thin jointed structures rather like the legs of an insect, they are quite different from the leaf-like type common in some of the previous crustacean groups. The full list of limbs is as follows : two pairs of antennæ, four pairs of jaws, seven pairs of thin thoracic limbs, six pairs of abdominal limbs of which the first five are broad plates (gills) which are used for breathing. The head is small and the eyes do not have stalks ; the thorax is very clearly segmented, but the abdomen is more fused together and spreads out at the back as a broad flat plate underneath which are found the gills.

The females lay their eggs about April or May. After being laid they are carried about by the female, individuals with eggs being easily recognised because they have a large white swelling near the head end on the ventral surface. This white mass consists of a mass of eggs which are held in position by thin flat plates which are developed from the bases of the front four pairs of walking legs. The plates which hold the eggs in position are so transparent that the embryos in the eggs may easily be seen through them. The eggs hatch when the young are at a fairly early stage of development, and the young remain inside their protective covering until they have developed further. If you wish to get a supply of young water-lice, females carrying eggs or young should be isolated in a fairly small dish and observations made daily until the young are set free. They are then very like their parents, but they are completely transparent because they have no pigment in their skin and they have the last pair of legs still undeveloped. Such young water-lice are one of the best objects for observing the circulation of the blood in Crustacea ; the individual blood corpuscles can be seen circulating along the limbs, antennæ and in the gill plates ; the long dorsal heart stands out beautifully and its rhythmic contractions can be watched. The circulation is normally obscured in the adults by the development of the pigment in the skin.

Amphipod Crustacea.—This is the group of Crustacea to which belong the familiar sea-shore " sandhoppers." There is one very common species, the so-called Fresh-Water Shrimp (*Gammarus pulex*), which is to be found in many of the rivers and streams of this country (see Fig. 67). They seem to prefer running water which is shallow and therefore probably containing much oxygen ; they are also to be found in small shallow lakes which have a stream flowing in and out. Three other species of *Gammarus* occur in the slightly salt-water at the mouths of rivers. They are *G. chevreuxi, G. duebeni,* and *G. zaddachi;* with the exception of *G. duebeni* they are not found in completely fresh water.*

The general shape of a fresh-water shrimp is shown in Fig. 67. The body is much flattened from side to side, and the outline is like the arc of a circle. The animal swims about on its side, and as it moves the hind third of the body is

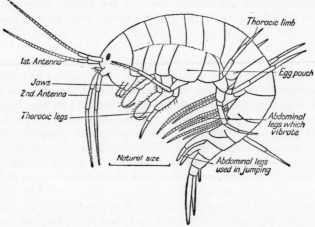

FIG. 67.—The Fresh-Water Shrimp (*Gammarus pulex*), side view.

straightened into the arc of a circle with a larger radius, then contracted suddenly ; the limbs help in swimming, but when they are at rest at least three pairs of thoracic limbs in the middle of the body always take up the curious position shown in the figure. (They are reflexed backwards.) The first three limbs on the abdomen vibrate continually and they produce a current of water over the gills at the base of the thoracic legs. The following three pairs of abdominal legs are curved backwards and are used for jumping. The body colour ranges from a pale reddish-brown to a grey-brown and the size goes up to about 2 or 3 cm. in length. They

* Species with a very local distribution are, G. lacustris, found in Lochs of the Inner Hebrides, Caithness, Sutherland and Ross-shire ; *G. sarsi* found in Loch Stennis, Orkney ; and *G. trigrinus* occurring in the saline waters of Cheshire, Droitwich in Worcestershire, and Gloucestershire.

live normally under stones in a stream bed where they can get some protection from the current. When you move the stones under which they are taking cover they swim away a short distance and then take cover under another stone. In the small streams in the counties north of London these creatures may be very common indeed, so much so that the bottom of the stream is almost moving with them.

Fresh-water shrimps do not usually live more than a few days in an aquarium unless you keep the water ærated by passing bubbles of air through it. They are quite unlike the water-louse (see p. 105) which will live for months in a small jar of water. In spring and summer when the females are carrying eggs or young in their egg pouches they are themselves carried about by the males. When the young hatch from the eggs they have all their legs so that there is no larval stage. Young specimens are of course small, their eyes are also different. The adults have eyes which are nearly always black, and which are made up of a large number of separate little black spots ; in young individuals these spots are fewer, and the colour is a reddish-brown rather than black. Adult specimens with bright red eyes are found occasionally, this factor for red eyes being inherited on Mendelian lines.

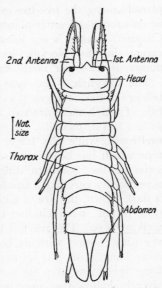

FIG. 68.—*Corophium lacustræ*, drawn from a preserved specimen from the Norfolk Broads.

Another Amphipod which is common in the slightly salt-water of some of the Norfolk Broads (River Thune and Heigham sound) is shown in Fig. 68. This animal is called *Corophium lacustræ ;* it is smaller than the fresh-water shrimp, and the body is not compressed from side to side. The most obvious feature is the pair of very stout second antennæ which are carried bent double. The abdomen is shorter than in the shrimp and it is held straight out, not curved round. The colour is greyish and *Corophium* usually lives among colonies of the cœlenterate *Cordylophora* (see p. 16) where its colour matches its surroundings. Quite a small amount of *Cordylophora* may harbour twenty or more specimens. Each individual constructs a tube of mud, which is either placed among the tree-like growth of the *Cordylophora* colony, or on submerged water-plants ; they feed on organic débris. *Corophium lacustræ* is not known to occur anywhere in Britain except in the Norfolk localities mentioned. There are numerous sea-water members of the genus.*

The Crayfish.—The only fresh-water crustacean which belongs to the same group as the lobster and crab is the crayfish, of which there are two species in this country : *Astacus pallipes* and *A. fluviatilis*. *A. pallipes* is thought to be indigenous to Britain, while the larger *A. fluviatilis* is the continental type which has been imported and has escaped to colonise some of the Southern English rivers. A crayfish looks superficially very like a small lobster (Fig. 69). In colour it is dark brown, the size of adults varying from 4 cm. to about 9 or 10 cm., not including the pincers on the first pair of walking legs. The crayfish lives only in rivers, and most of the time it stays with its body protected by a hole in the river bank, with just its antennæ sticking out. Sometimes large specimens may be seen in the open stream. If the river is shallow they may be fairly easily captured with the hands if care is taken. They are still common in the rivers of Kent and Hertfordshire, also in Derwent, and they probably occur abundantly in many other places. I have found them under large stones which have been used to dam up part of a stream to make a fishing-pool, or inside old tins in a river.

In the crayfish the head is fused with the thorax to form the cephalothorax, a large unjointed part of the animal's body as seen from above. Behind this comes the abdomen, which is very clearly divided into segments, and which is very

* To identify British species of *Gammarus* see
Gammaridæ (Amphipoda) with key to the families of British Gammaridæ.
D. M. Reid (1944). Linnean Society of London. Synopsis of the British Fauna, No. 3.

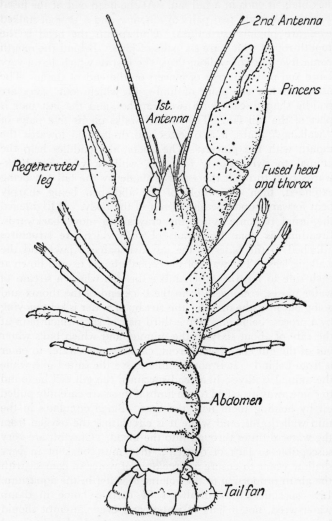

FIG. 69. — The Crayfish (*Astacus fluviatilis*), natural size.

flexible ; it ends in a tail fan. At the head end of the fused part of the body two pairs of antennæ and a pair of stalked eyes are obvious structures. Underneath the head is the mouth round which are six pairs of jaws. Behind the mouth come five pairs of walking legs, the first of which has a very large and powerful pair of pincers on the end of them. The abdomen has six pairs of limbs, of which four pairs are paddle shaped, each with two branches, and the last pair is part of the tail fan. The animal walks on its five pairs of " walking " limbs, it also seizes and drags food towards the mouth with the pincers. The abdominal paddles help the animal to swim slowly forwards, while a very fast backwards rush through the water is made by the tail fan being extended, then the whole of the abdomen being sharply bent forwards underneath the rest of the body. If frightened the usual response is for the crayfish to move backwards. Breathing is done by means of gills which are structures attached to the bases of the last two pairs of jaws and the walking legs. The walls of the cephalothorax extend down at each side to form a gill chamber through which a stream of water passes, coming in from the back end of the thorax and going out near the mouth ; this stream of water is kept moving by a process belonging to the third pair of jaws which flaps at the rate of about sixty to the minute, and which bails water out of the gill chamber in front, causing more water to enter it from behind. It is not possible to see the tufted gills while the animal is alive ; in a dead specimen the gill will be found to come away attached to the limbs if these are carefully pulled off by holding them near the base. Blood circulates in the thin walled gills, and while it is circulating the oxygen from the water diffuses through into the blood. Crayfish are very susceptible to lack of oxygen ; they must be kept in very shallow water to ensure a large area of water in contact with the air in proportion to the volume of water in the aquarium. Live specimens should always be carried home in damp water-weed, not in jars of water, and every attempt should be made to keep them cool during transit. If they live well for the first two or three days then they will probably live for months, but some nearly always die shortly after their capture.

Crayfish are carnivorous, eating any live animal of suitable size.   Prey is usually caught by the pincers which tear it into small pieces if it is too large to be put into the mouth at once. When the pincers have dragged the food towards the mouth it is further ground up by the jaws.   In captivity earthworms are readily eaten.

The males seem to be larger than the females ; they have their first pair of abdominal limbs made into structures looking like small tubes, which are used for transferring sperm to the female.   The females lay eggs in November, and from then on into late spring they may be found with a large mass

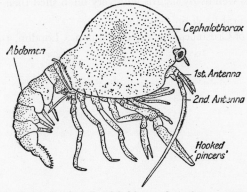

Fig. 70.—Newly hatched Crayfish (*Astacus fluviatilis*).   Modified after Calman.

of dark pink eggs attached to the hairs on the abdominal legs.   When they are nearly ready to hatch the eyes of the young show through the shell as two dark spots.   When the young hatch they remain clinging on to the limbs of the mother for some time ; they are fairly like their parents to look at, except that the cephalothorax is globular because it contains yolk from the egg ; the pincers have hooked tips for obtaining a firm hold of the female, and the first pair of swimming legs and tail fan are not well developed (Fig. 70). They are also carnivorous, eating of course much smaller animals.   They can be reared on a diet of small water-fleas. After the first moult the hooked tips of the pincers are lost and the tail fan appears ; at this stage the young are set free from the parent.

It is quite common to find a crayfish which has one pair of pincers much smaller than the other (Fig. 69), or one walking leg smaller than the rest. This is because it has been damaged at some time, and after such an accident the limb is shed and a new one grown in its place. The new limb is smaller than the others when it begins to grow, but it increases in size each time the animal moults. The power of being able to shed a limb, or the remains of a limb at will, is called *autotomy*, and it is possessed by other Crustacea such as crabs and lobsters as well as by lizards who very often shed their tails.

For a general book on Crustacea see
The Life of Crustacea, by W. T. Calman. London, 1911.

CHAPTER 8

THE ARTHROPODA : INSECTS

THE number of different kinds of insects in the world far
outnumbers all the other species of animals and plants put
together. Of this huge class of animals quite a few spend
part of their lives in fresh water, while a much smaller number
are aquatic all their lives. Many of the insect inhabitants
of ponds and streams are very common, and are likely to be
among the first living things to be observed by anyone beginning to study natural history. Many of them are large
enough when adult to be studied without the aid of a microscope, though a hand lens will be useful even for them, in
making out the details of their structure.

Insects are very easy to recognise in their fully grown (adult)
state. All have clearly segmented bodies which are divided
into three parts called the head, the thorax, and the abdomen.
The head carries a pair of antennæ or feelers used for smelling and touching objects surrounding the animal, though
sometimes the antennæ are so small that they are not readily
seen, as for instance in the caterpillar stage of butterflies or
moths. The head also possesses three pairs of jaws which
are really greatly modified and shortened legs clustered
round the mouth. They are subject to great variation in
shape according to the diet consumed, and a knowledge of
the type of mouth-parts which characterise the insect groups
is of much help in identifying specimens. In general these
mouth-parts consist of a pair of strong biting or chewing jaws
(the *mandibles*), a pair of accessory jaws behind them provided with sensory processes for testing the food (the *maxillæ*,
with *maxillary palps*), and a third pair which are always

joined together, forming a kind of lower lip (the *labium*).
The latter are also provided with sensory processes (the
labial palps). The jaws of a beetle are illustrated in Fig. 71
which will give some idea of the appearance of the various
parts in an insect with mouth-parts of the biting type.
Typically the head of the adult insect carries two pairs of
" compound " eyes, with in some cases three dorsally placed
simple eyes as well. A compound eye has a large number
of separate lenses which each produce an image of the objects
immediately in front of them. This " mosaic " of separate
images is somehow resolved into a picture of the outside world.

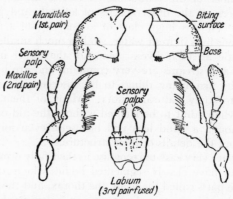

Fig. 71.—Mouth-parts (jaws) of *Gyrinus*, the Whirligig Beetle.

The simple eyes or ocelli contain groups of pigmented cells
which are sensitive to light.

The thorax consists of three segments each bearing a pair
of jointed legs. The principal joints of the leg are constant
in number throughout the adult stages of all insects, though
the individual joints may be modified in shape for special
functions such as swimming, digging, and so on. The type
of leg used for running or walking best represents the basic
form from which the others are derived. It has five distinct
parts as follows (see Fig. 92) : a small stout portion attached
to the underside of the thoracic segment—the *coxa*, followed
by another small joint—the *trochanter* ; next come the two

longest parts of the leg, the *femur* and the *tibia*. Attached to the distal end of the tibia is the foot part of the leg or *tarsus*, which may be made up of from one to five small joints, the last one usually bearing claws. The number of joints in the tarsus is an important clue to the identity of some insects.

The number of walking legs is so constant (i.e. six) that it is safe to assume that any segmented creature which you find in the water which has *six* legs is an insect. In the winged forms the second and third segments of the thorax have each a pair of wings ; the first pair in the bugs and beetles are hardened structures used for protecting the second thin membranous pair with which the flying is done. Many beetles appear to have no wings at all until you gently displace the protecting pair to one side and find that they were making a false back to the insect. Nearly all wings are strengthened by ridged " veins " which have a characteristic arrangement, the *venation*, in the different orders. In the group of insects called the true flies (Diptera—the group to which the housefly, bluebottle, and mosquito belong) the second pair of wings is reduced to small knobs which are used as balancing organs. Flying in the Diptera is done with the first pair only.

The abdomen is made up of distinct segments varying in number from eleven in dragon-flies to six visible ones in flies. There are no legs on the abdomen of any aquatic insect, though certain very small paired structures near the anus are thought to represent abdominal legs ; there may also be long tail processes (*cerci*) attached to the last segment, and gill structures on all or some of the other abdominal segments.

The bodies of all insects, as of other arthropods, are covered by a material called *chitin*. This substance is secreted by the living skin cells lying underneath, and certain internal structures are also covered by a layer of it, as for instance the fore and hind part of the gut and the inside of the breathing tubes (*trachea*) to be mentioned later. Chitin in thin layers is extensible ; but in thicker layers and when impregnated with other substances it becomes hard, brittle, and impermeable. In such cases it forms an external skeleton supporting the soft tissues inside. There is some internal skeleton as well, because ingrowths of chitin at certain places on the body provide struts for the attachment of muscles. In young stages a chitinous external covering involves difficulties concerned with growth, hence the skin has to be moulted at intervals to allow for increase in size.

Since many of the insect forms found in fresh water are young stages, and since they are often different from the adults to which they belong, it is convenient here to describe in general terms the young insect. Some young stages closely resemble the adult, as do those of wingless insects, where the only difference is small size and sexual immaturity of the young. These are called *nymphs*. In some of the winged insects, the young stages also resemble the adults, but wings are not present until after several moults, and when they do appear they project like short plates from the back of the thorax ; they are not functional. Compound eyes characterise these active forms which are also called nymphs. Other young stages bear no resemblance to the adults. Their developing wings never show on the outside, but develop internally, while their legs are often short, stumpy structures with less sign of jointing than those of the adults, or they are absent altogether. Their eyes are the simple pigment spots known as *ocelli*. Compound eyes are never present. These young stages, in which the developing wings are never visible externally, are called *larvæ*. They are found in beetles, alder-flies, caddis-flies, moths, Hymenoptera (wasps, bees, ants, etc.), and the Diptera (true flies). Between the larval and the adult stage in these insect groups, another, usually dormant, phase occurs in the life-history, in which the developing wings appear on the surface for the first time ; this is known as the *pupa* or *chrysalis*. In fresh-water both types of young insect, nymphs and larvæ, abound ; the less obvious pupæ may often be found in the mud at the pond side, to which place the larvæ repair at the time for pupation. Pupæ may also be found floating passively on the water surface (as in the rat-tailed larva, *Eristalis*). In some cases the pupa is active, as in many flies, including mosquitoes ; here movements of the tail-fan cause them to swim downwards ; when the swimming movements cease they float automatically up to the surface. The caddis pupa possesses a pair of beautiful swimming legs with which it swims to the surface, or the water's edge, before emerging as an adult.

From this brief description of the young insect stages it will appear obvious that these animals can be divided into two groups in one of which the young are nymphs resembling in general form the adults ; development in this case is said to be direct. The aquatic insects in this division are the springtails, dragon-flies, stone-flies, may-flies, and bugs. The other group go through larval and pupal stages before the adult form is completed ; these are said to have indirect development. The aquatic insects in this second division are the alder-flies, beetles, moths, Hymenoptera, caddis-flies and true flies. This latter kind of life-history is more complicated than the former, and more difficulty will be experienced in identifying the young stages of members of this group.

Among the nymphs of certain insects whose adults do not live an underwater life, extra structures such as gills are developed, which are generally lost at the final moult so that they are not present in the adult. This is the case in the nymphs of dragon-flies, stone-flies, and may-flies ; the stone-flies, however, may retain the nymphal gills as vestiges.

The whole group of insects is organised in relation to life in air. This is best seen in their breathing organs which consist of a much branched net-work of tubes opening to the exterior at definite holes in the skin called *spiracles* or *stigmata*. Internally the branches of this tube system supply all the organs with a direct supply of air from which oxygen is removed. The insect's blood plays little part in respiration. These air-containing tubes, or *tracheæ*, are supported internally by thin layers of chitin which prevent their walls from collapsing together. When tracheæ are viewed in the natural state under a hand lens they appear like silvery threads (on account of their contained air) becoming smaller as they branch and pass to their destinations. They are only visible when the skin is transparent as it is in many aquatic larvæ.

In the adults of many aquatic insects, such as the beetles and bugs, the spiracles remain open, and they communicate with a store of air which is carried down against their bodies under water. Thus in the diving beetle (p. 156) a store of air is carried in a space between the wing-covers and the back ; this is utilised under water and replaced with fresh air by the beetle coming to the surface (see p. 155 for further information). In the water boatman (*Notonecta*) the hairs on the underside of the body are unwettable. They form an archway beneath which lies a column of air which is in contact with the open stigmata on the thorax. This entrapped air gives the animal a silvery appearance as well as making the animal so buoyant that it automatically floats up to the surface if it is not swimming or holding on to some submerged object.

In a large number of water insects the breathing system is closed because the spiracles do not open on the surface. These are unable to breathe air directly and so do not appear at the surface for this purpose. They have instead special breathing processes, richly supplied with tracheæ, which are called tracheal gills. Through them oxygen is taken up from the surrounding water, and carbon dioxide given off. The form and position of tracheal gills varies greatly. They may be slender, whitish filaments as in caddis larvæ, or plate-like or feather-like appendages on the sides of the abdomen as in may-fly nymphs. Certain dragon-fly nymphs have this type of gill developed in the wall of the hind part of the gut and they pump a current of water over these from which oxygen is obtained.

All insect eggs, except those of a few parasites, are provided with much food material and are laid enclosed by a shell which is often hard. In the fresh-water forms the eggs are sometimes laid by the female flying over the water surface and dropping them at random. More usually they are laid singly or in groups on plants either above or below the water surface. In some cases they are laid surrounded by mucilage which swells up on contact with water into a jelly mass. Some mosquitoes lay a large number of eggs arranged vertically like a bundle of cigars on the water forming an egg-raft which floats on the surface film.

In the following pages the various orders of aquatic insects are dealt with individually.

Springtails (*Collembola*).—There are two small species of springtail which live on the surface of fresh water. They are called springtails because they spring into the air by means of a forked tail process which is kept tucked up on the underside of the body when the animal is at rest. Much the more common of the two species in the black *Podura aquatica* (see Fig. 72) ; it may be found in large numbers on the surface of any kind of still water. (The water looks as though it were covered with specks of soot.) The body is a dusky black, and an individual appears as an elongated speck to the naked eye, but as they are usually found in large numbers together they are

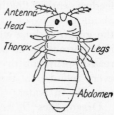

FIG. 72.—The Spring Tail (*Podura*), length 1½ mm.

more conspicuous than you might imagine from their size. They are sufficiently small for all their weight to be supported by the surface film of the water ; they seldom go below the surface, and their skin is not wetted by the water at all. They die if removed from the water and placed in a dry box, because they are unable to live in dry air as this causes the water inside their bodies to be evaporated. They can only live in damp surroundings. The general shape of the body is shown in Fig. 72, but to see the tail process the insect must be placed

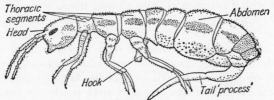

FIG. 73.—The Spring Tail (*Isotoma palustris*), length about 1 mm. After Miall.

on its back. This appendage is held in place along the underside by means of a kind of hook ; when the tail process is freed from the latter it hits the surface of the water with sufficient force to make the animal leap into the air.

The other species of aquatic springtail is *Isotoma palustris* (see Fig. 73), which is smaller than the black *Podura* and has much more conspicuous antennæ. It is less common.

Young springtails are just like the adults except for their sexual immaturity and their size. All are wingless and many breathe through the surface of their bodies instead of through tracheæ as in other insects. They are apparently vegetable feeders, eating floating plants, but little seems to be known about their exact food. As breathing in many members of the group is done entirely through the skin, this means that it is permeable to water vapour as well as to air. For this reason springtails dry up and die quickly in dry air ; in nature every member of the group lives in a damp situation, if not actually in water.

Dragon-flies (*Odonata*).—The adult winged dragon-fly is a very familiar object to anyone who has walked or rested in a sunny glade near a stream or pond in summer. The large species, such as *Cordulegaster boltonii* (Fig. 74A), are almost terrifying, they are so noisy on the wing, while the more delicate Damsel flies, with their banded and brightly coloured bodies, are very graceful in their movements. As they dart about in the air dragon-flies catch their food, which consists mainly of gnats and mosquitoes ; if a specimen is captured and examined it may be found to have a whole number of these insects in its jaws, and for this reason they are sometimes called " Mosquito Hawks." The two pairs of large wings are transparent with a net-work of veins all over them.

The young stages of all dragon-flies are completely aquatic, and for living in water they have various structures not possessed by the adults ; because of these they are rather unlike the adults in appearance and are sometimes called *naiads*. Species of dragon-flies with long bodies lay cylindrical eggs which the female places one at a time inside the tissues of some water-plant either above or below the water surface. Those with shorter, broader bodies scatter their eggs at random over the water surface. The eggs may hatch in a month or six weeks but it seems likely that they may often wait until the following spring. Almost immediately after hatching the young goes through its first moult. The length of time spent in the naiad stage varies with the species, but it seems to be from one to three years during which period the skin is moulted on about twelve occasions. If food is plentiful then the dragon-fly probably completes its development more quickly. By no means all the stages are known in the development of the British species, least being known about the youngest period of development, when it is difficult to identify the species. Somewhere around the fifth or sixth moult the wings begin to make their first appearance, and these increase in length at each subsequent moult. When the time for the final moult arrives the full-grown naiad climbs up a water-plant out of the water and rests for a short time. Its skin

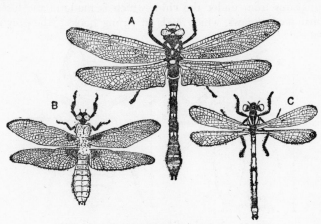

FIG. 74.—Adult Dragon-Flies, about ⅔ nat. size.
A. *Cordulegaster boltonii.*
B. *Libellula depressa.*
C. *Coenagrion puella.*

splits down the back of the thorax and the animal's head, thorax, and three pairs of legs are pulled out of the skin ; after this a rest usually occurs before the abdomen is extricated. The newly emerged adult has only small useless wings and a fairly short body, but these quickly expand and grow in length until they have both assumed the typical size and shape of the adult. The expansion of the wings is due to blood being forced into the veins, while the body expands by the creature swallowing air. In two to three hours or less, the change from full-grown naiad to the adult dragon-fly is completed.

Every dragon-fly naiad is carnivorous, and some of the largest kinds even eat tadpoles ! The food is captured in a most curious way, partly perhaps because the naiads are clumsy and slow at getting about. Their general body colour is always some dull shade of green, brown, or grey, varying in specimens of the same species to match the background of the particular pond where they were found. They are thus very inconspicuous while they remain still. If suitable prey comes fairly near, the naiad suddenly shoots forward a long apparatus from under the chin, which is made of the fused third pair of jaws, which ends in strong hooks ; the prey is

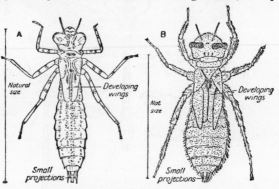

FIG. 75.—Nymphs of Anisoptera.   After Lucas.
A. *Æshna cyanea.*
B. *Libellula quadrimaculata.*

seized by the hooks, then the whole apparatus is pulled back bringing the food to the mouth. At rest this special structure is folded back under the head hiding the lower part of the animal's face ; for this reason it is often called " the mask " The mask enables the naiad to reach its food while it is still some distance away.

There are two different methods of breathing used by the naiads, and this serves to separate the species into two principal divisions. The first group have broad and rather fat bodies with a number (5-6) of very small projections at the hind end (Fig. 75). Water is pumped in and out of the anus, passing over gills which line the hind part of the alimentary canal. If water is forced in and out in rapid succession the insect is shot forwards in a series of jerks which is a quicker method of progression than by its prowling walk. This current of water from the anus can easily be seen if a specimen is placed in a glass dish containing water with some

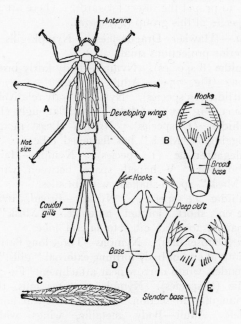

FIG. 76.—Nymphs of Zygoptera.
A. *Coenagrion puella*, the Damsel-fly.
B. Mask of same (much enlarged).
C. Gill filament of same.
D. Mask of *Agrion*.
E. Mask of *Lestes*.

mud particles and poked with an instrument to encourage it to move away rapidly. There are twenty-seven British species in this group (the *Anisoptera*).

In the second group all the naiads have long slender bodies with three long projections from the hind end of the body which are usually flattened from side to side. These contain a large number of tracheæ and are usually considered to act as gills ; the insects, however, are not dependent on them as they can still live when the "gills" have been accidentally lost. If such a gill is lost while the insect is still young a new one is grown in its place.   This type of naiad waves its body from side to side in a kind of swimming movement in which the gills help to propel the insect forwards.   There are seventeen British species of this group (*Zygoptera*).

Anisoptera.—(Hawker Dragon-flies).   Nymph-gills internal, posterior projections small.

Æschnidæ (8 species).   Nymph—body fairly broad, quite long (Fig. 75A).   Adult, Fig 74A.

Libellulidæ (13 species).   Nymph—body relatively fat and short (Fig 75B).   Adult, brown, flat, ugly (Fig. 74B).

Gomphidæ (1 species).   Nymph—Legs modified for digging.   "Mask" has flattened lobe.

Cordulegasteridæ (1 species).   Nymph—Hairy, oval abdomen.   V-shaped space between wing cases. "Mask" spoon-shaped, with bristles.

Corduliidæ (4 species).   Nymph—Broad, oval body, seven short, jointed antennæ.   "Mask" spoon-shaped, toothed edge to lateral lobes.

Zygoptera (Damsel flies).   Nymphs—Three long flattened projections at end of body forming external "gills."   Adults —slender, wings narrow near attachment (Fig 74C).

Agriidæ (2 species).   Nymph—2 outer gills, thick and triangular in section.   "Mask's" central joint deeply cleft.   Adult—Body metallic, wings wider than Cœnagrion.

Lestidæ (2 species).   Nymph—All 3 gills thin plates. Brownish.   Base of "Mask" bearing terminal hooks, very slender, and cleft at distal end.   Rare.

Platycnemididæ (1 species).   Nymph—short, broad. "Mask" has projecting median lobe, no cleft. Caudal gills thickened leaflike appendages, ending in a long slender curved point.

Coenagriidæ (12 species).   Nymph—green or brown. "Mask" base bearing terminal hooks, quite broad. Not cleft at distal end (Fig 76A, B, C).

For further information on Dragonflies see page 288.

Stone-flies—(*Plecoptera*).   Adult stone-flies are only found near water.   They are brownish coloured insects with very long antennæ and usually two long tails at the hind end. Some species are large, measuring 5 cm. or more across their wing span, but all are bad fliers in spite of their two similar pairs of large wings. In the males the wings may be reduced in size so as to be functionless. These insects are short lived in the adult stage, but are not so helpless as is commonly believed. All drink liquids, and most will feed on lichens and algæ. The females drop their eggs in one mass into the water ; there they separate, each becoming attached to some submerged object by means of a long thread which it possesses at one end. In the genus *Perla*, to which the largest species belong, the eggs are black and oval shaped ; they may be seen projecting in a bundle from the hind end of the female's body before she drops them into the water.

The eggs, which are laid in summer, hatch in about six or eight weeks, and the small newly hatched nymphs shelter from water currents among moss and plants. The nymphal life lasts for a year or more, and most stone-fly nymphs are found in small quickly flowing mountain or moorland streams. In winter the nymphs often hide in mud, but they reappear in spring. From then on they grow rapidly. They are easily recognised by their two tails, two clawed feet and ten abdominal segments (Fig. 77). The older nymphs cling to the undersides of flat stones which are in the quickest flowing part of the water. Those of the bigger species look something like large earwigs at first sight. The aquatic stages of stone-flies are found commonly in the same places and on the undersides of the same stones as some may-fly nymphs ; they may be distinguished from may-flies by the two tails at the posterior end of the body. (May-fly nymphs with one exception, have three tails.) The nymphs are not good swimmers ; they can crawl about on stones pressing their flattened bodies close against the surface in strong currents or they can walk along the stream-bed with a rather clumsy gait swaying their abdomen from side to side all the time to help themselves to go forward. When you find them on the underside of a wet stone you will soon recognise them from the way they wriggle along. In captivity nymphs of the larger species always sway their body up and down rhythmically, a movement which they may also do in the stream, and which would appear to help respiration through the tubular tracheal gills. In an aquarium they do not live well unless provided with stones under which they can crawl to avoid light.

The youngest stone-fly nymphs have no developing wings or gills on their bodies ; breathing is done through the skin. In small species no special gills are ever developed, while some larger species grow tufts of tubular tracheal gills on the thorax and the end of the abdomen, after several moults (see Fig. 77). In all species wings begin to appear after a number of moults, and when you find a specimen in which these are fairly long, then you know that it is nearly ready for its last moult, after which it will become adult and leave the water. Some nymphs of the larger species appear to leave the water periodically for several hours. They usually do this at night and they moult out of the water, leaving their cast skin on the plants and rocks at the water's edge before returning to the stream.

Some stone-fly nymphs are carnivorous, eating small worms, insects and Protozoa when they are young ; when they grow larger, may-fly nymphs, dragon-fly naiads, and caddis worms seem to be their chief diet. Other nymphs are vegetable feeders, or are able to ingest detritus of animal or vegetable origin.

When a nymph is full grown and ready for its final moult it climbs out of the water on to a rock or other convenient surface. The insect then blows up its body by swallowing air. The final moult takes place rapidly and the insect emerges from its last nymphal skin as an adult.

Six families of stone-flies are represented in Britian :—

Family Perlidæ

Three genera, *Perla*, *Isoperla* and *Chloroperla*. The nymphs of *Perla* are the largest and have the most robust-looking bodies of any stone-flies (see Fig. 77). Tubular gills are present on the thorax and the end of the abdomen in all except the very youngest stages. The largest species *Perla carlukiana* (called *P. maxima* in many books) is not uncommon in hilly streams ; the adults have two prominent cerci like the nymphs. The nymphs of the genus *Isoperla* (Fig. 78) are more slender, and the greenish adults, with a wing span around 20 mm., are smaller. Of the two British species *Isoperla grammatica* is common in almost any fairly swift flowing stream, while *I. griseipennis* is rare. *Chloroperla* is the third genus ; these stone-flies resemble *Isoperla*, but are smaller still, and greenish in the adult stage. The species *C. torrentium* is common.

Family Nemouridæ

One genus, *Nemoura* containing ten British species. The nymphs

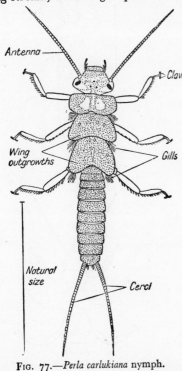

FIG. 77.—*Perla carlukiana* nymph.

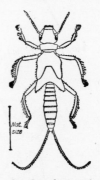

FIG. 78.—*Isoperla* nymph. After Rousseau.

(Fig. 79) are common and may be found in running water and still water. They frequently live among water filants and are to be found among liverworts and mosses which are covered by a small waterfall. The adults have rudimentary cerci.

Family Tæniopterygidæ

The four British species belong to three genera, *Rhabdiopteryx*, *Nephelopteryx*, and *Tæniopteryx*. *T. risi* (Fig. 81) is very common.

9

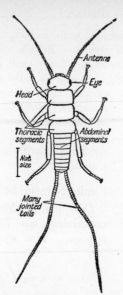

FIG. 79 *Nemoura*, young nymph.

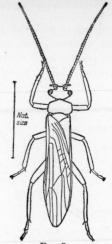

FIG. 80.
Nemoura, newly emerged adult.

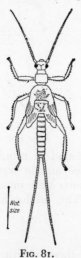

FIG. 81.
Tæniopteryx nymph. After Rousseau.

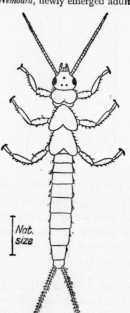

FIG. 82.—*Leuctra* nymph.

Family Leuctridæ

One genus, *Leuctra*, containing six species. These nymphs are small and fragile, and are commonly found under stones in swift streams. They also burrow in sand, and die in captivity if exposed to too much light.

Family Capniidæ

One genus, *Capnia*, with three British species, only one of which, *C. nigra*, is fairly common in running water and near the shores of lakes. The nymphs are small and slender, rather like *Leuctra*, but with shorter cerci and antennæ.

Family Perlodidæ

Two British species, belonging to different genera, that is, *Perlodes mortoni* and *Dictyopterygella bicaudata*. Both are not uncommon in the Lake District.

For further reference on this group the following papers may be consulted :—

1. A key to the British species of *Plecopters* (Stone-flies), by H. B. N. Hynes (1940). Fresh-water Biological Association of the British Empire, Scientific Publication No. 2.
2. The taxonomy and ecology of the nymphs of British Plecoptera, by H. B. N. Hynes (1941). Transactions of the Royal Entomological Society of London. Vol. 91, part 10, p. 459.
3. Handbooks for the Identification of British Insects. *Plecoptera*, by D. E. Kimmins (1950), Royal Entomological Society, London.

May-flies (Ephemeroptera).—Adult may-flies like stone-flies are also found only near fresh water. They have an extremely short adult life lasting in some cases only a few hours, in others two or three days. The young stages (nymphs) are aquatic, and it may take one or two years from the time of hatching from the egg for the insects to reach the adult stage. On fine summer days the adults emerge from the water in large numbers, and it is then and later during their nuptial flights, that they are eaten extensively by fish, particularly trout. Some artificial " flies " used by trout fishers are models of various species of may-fly.

The nymphal stages of may-flies are easily recognised though they exhibit some diversity of shape. The very young nymph has no gills, but half-grown nymphs have a series of tracheal gills attached to the sides of the abdominal region, which are differently shaped in the various genera, and which therefore help in identifying specimens ; there are three " tails " at the hind end, and the tarsal part of each leg has only one joint ending in a single claw (see Figs. 84, 89, 92). The nymphs really pass through four slightly different stages during their development ; in each stage they moult many times. The newly hatched ones breathe through their skin, but after about ten days, when they are about 1 mm. long, the gills begin to grow. Slightly older nymphs (second stage) have

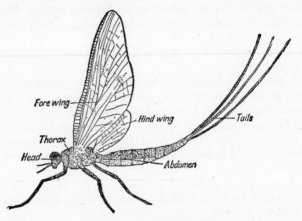

Fore wing

Hind wing

Tails

Thorax

Head

Abdomen

Fig. 83.—An adult May-Fly.

well-developed gills. In the third stage the wing outgrowths begin to appear on the thorax in addition to the abdominal gills ; this is the commonest stage in which to find may-fly nymphs, because they are by this time fairly large. Before the penultimate moult, when the third stage is full grown, the insects come up to the water surface and go through a very quick moult after which they fly up into the air. In this fourth stage the insect has lost its gills, the wings are nearly full size and are functional, but it has one more moult to go through. May-flies are the only insects to have a moult after they have reached the adult form with the power to fly. This period in the insect's life is usually short (it is called the sub-imago), and it is in this state that the insects " rise " from the surface of the water into the air. They settle on some object near the water to shed their final skin, from which the true adult comes out. Both the adult and the sub-imago have three long tails on the abdomen (Fig. 83). After the females have been fertilised they lay their eggs in the water, usually by depositing them on the surface, but in some genera, such as *Bœtis*, the female descends underneath the water and places her eggs on submerged stones, carrying with her under her roof-like wings a volume of air which she breathes.

The nymphs bear a strong resemblance to the kind of adult into which they will develop. Each genus has a more or less typical form of nymph, but whereas the various species are well known in their adult stage, the species of the nymphs are only known in comparatively few cases. To be sure of the species of a nymph it is necessary to rear it through to the adult, and this is usually difficult because they easily die. They are usually vegetable feeders, most of them living on algæ which they rasp off stony surfaces. The gills on the abdomen undergo rhythmic movements in life. There are about forty species of British may-flies belonging to thirteen different genera. A typical nymph of each genus is briefly described below with any interesting note about its habits. They are divided into four types which depend on their mode of living. They are (*a*) Burrowing nymphs ; (*b*) Much flattened nymphs which live in strong currents ; (*c*) Swimming nymphs ; and (*d*) Creeping nymphs.

(*a*) *Burrowing Nymphs.*—The only may-fly nymphs in this country which are really adapted for burrowing belong to the genus *Ephemera*, of which there are perhaps three British species. These nymphs live in tunnels which they make in sandy mud near the edge of streams and small rivers. They are common, but are often overlooked because they usually only become visible when the mud is scooped up on to the bank and examined. Occasionally they swim in muddy water if the light is not too intense. The nymphs are very unhappy if placed in clear water with no mud, and they soon die in captivity if they are not provided with material in which to burrow.

Here again, as is the case with the stone-fly *Leuctra*, too much light is detrimental. The general shape of the nymph is

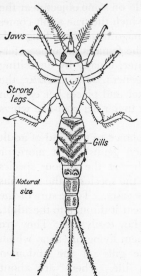

seen in Fig. 84. When nearly full grown they are 3½ cm. long ; the body is elongated ; at the head end one pair of jaws (the mandibles) is very long and strong for digging. The legs are also strong and adapted for digging ; they are very hairy. The feathery gills on the abdomen are reflexed over the back. The nymphs eat mud, living on the organic matter, chiefly Diatom

FIG. 84.—*Ephemera* nymph.

plants, which it contains ; they only leave their burrows if the level of the water falls, as they would die if the mud dried ; or because they have grown too big for a particular burrow and have to construct a new one. The adult female lays thousands of eggs which are scattered over the water surface.

Nymphs of the genus *Potamanthus* are somewhat similar to *Ephemera*

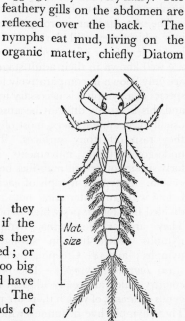

FIG. 85.
Potamanthus luteus nymph.
After Eaton.

nymphs, but they have lost the burrowing habit and shelter under stones at the edge of streams. There is one British species *P. luteus* (Fig. 85).

(b) *Flat Nymphs living in strong currents.*—A number of genera come under this heading, figures of which are shown on page 137. They all have some degree of flattening of the body and of the limbs, while the claws on the feet are particularly strong for clinging to stones in fast-flowing water.

Genus Habrophlebia is really intermediate between the burrowing and the flattened type of nymph. One species, *H. fusca* (Fig. 86), is found in small streams in this country.

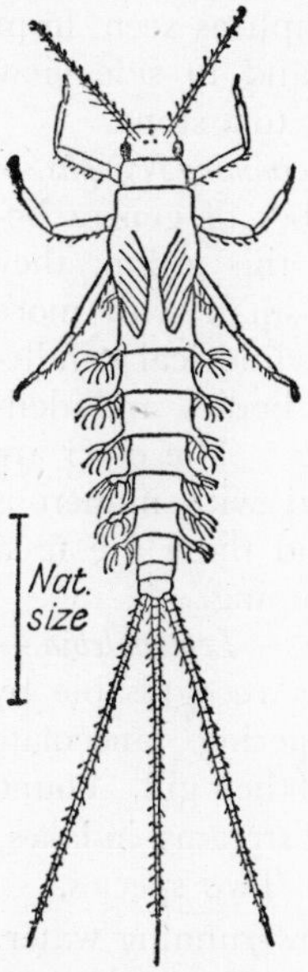

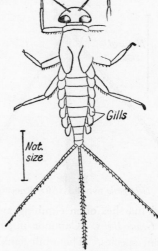

Fig. 86.—*Habrophlebia fusca* nymph. After Eaton.

Fig. 87.
Rhithrogena nymph. After Eaton.

Genus Rhithrogena.—Two species, *R. semicolorata* and *R. haarupi*, found in Britain. The nymph (see Fig. 87) is fairly flattened, and it is found clinging to the undersides of stones in swift-flowing streams and rivers.

Genus Heptagenia.—The nymph is fairly flattened and has large flat limbs ; the cerci are very long and the gills are feathery (see Fig. 88) Four species.

Genus Ecdyonurus.—The nymph is extremely broad and flat with very large flat limbs. There are four species. The nymphs are found in swift-flowing water, chiefly mountain streams and at the edges of large lakes. They can maintain a hold of the stone in very swift currents by means of their strong clawed feet. Figs. 89, 90 show the nymph as seen from above, and in side view clinging to a stone.

(*c*) *Swimming Nymphs.*—A number of genera belong to this group, they are all small with more or less cylindrical spindle-shaped bodies and delicate legs. The cerci are provided with numerous hairs and these are used in swimming.

Genus Leptophlebia.—Nymphs recognisable by the tapering lanceolate shape of their gill. Found in small streams, in lakes, and marshes often climbing among plants. Two species.

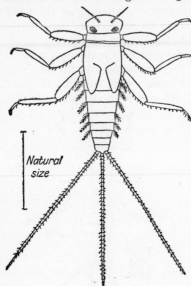

Natural size

FIG. 88.
Heptagenia nymph. After Rousseau.

Genus Bætis.—Common in streams and slow-running water. Nine species. The adult female descends under the water to lay her eggs on stones (Fig. 91).

Genus Centroptilum.—Very similar to *Bætis* but with longer antennæ. Found in streams. Two species.

Genus Cloëon.—Very active small nymphs found in ponds, ditches, and lakes. They are very common and exhibit really beautiful swimming movements by means of the abdomen and hairy cerci. Two species (Fig. 92).

Genus Siphlonurus. Found chiefly in hilly streams and mountain tarns ; a good swimmer, it lives under stones or among aquatic plants. Two species (Fig. 93).

(*d*) *Creeping Nymphs.*—Two British genera which neither

swim nor burrow much, but just move slowly along on the surface, or slightly embed themselves in the sand or mud.

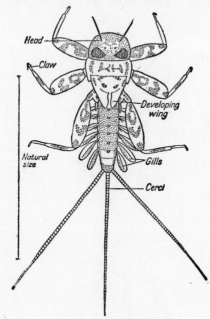

FIG. 89.—*Ecdyonurus* nymph.

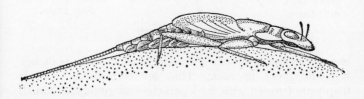

FIG. 90.—Nymph of *Ecdyonurus venosus*, clinging to the surface of a stone. After Eastham.

Genus Cænis.—The nymphs live among mud, water-plants, or algal threads in still water. They are very easily recognised on account of their curious gills. The first pair are vestigial, the second pair form a hard covering which protects the remaining four pairs of gills which are arranged in series underneath them. The nymphs are fairly sluggish, but at least

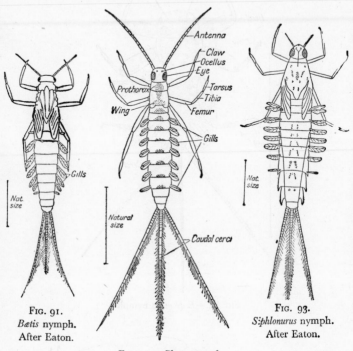

FIG. 91.
Bætis nymph.
After Eaton.

FIG. 92.—*Clæon* nymph.

FIG. 93.
Siphlonurus nymph.
After Eaton.

one species can swim with a movement which looks as though it would break its back. They are carnivorous and are frequently covered with mud particles and epiphytic animals which perhaps serve as a kind of camouflage when they are stalking their prey. *Cænis horaria* (see Fig. 94) burrows slightly in mud, but not far enough to cover the gills. Three British species.

Genus Ephemerella.—Nymphs are found creeping along the mud on the bottom of river and stream beds. Like *Cænis* they cover themselves with débris, but they are easily distinguished from the latter by the gills (see Fig. 95). Two species.

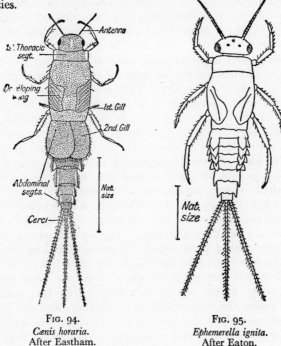

FIG. 94.
Cænis horaria.
After Eastham.

FIG. 95.
Ephemerella ignita.
After Eaton.

To identify genera of Nymphs see
Biologie der Süsswasserinsekten, by von C. Wesenberg-Lund.
    Berlin and Wien, 1943.
Key to British Species of Ephemeroptera, 1942, by D. E.
    Kimmins. Freshwater Biological Association of the
    British Empire, No. 7 (1942).
For further information on Nymphs see page 288.

The Water-bugs (Hemiptera).—Many of the larger water-bugs are among the most familiar inhabitants of ponds because they are exceedingly common, they are large enough to be easily seen, and they may be kept very successfully in captivity. The order of insects known as bugs (*Hemiptera*) contains many terrestrial forms which do much damage to man's crops, such as the leaf-hoppers, white-flies, aphids, scale insects, and mealy-bugs, but most water-bugs are harmless to plants because they are carnivorous and live on the live or dead bodies of other animals. The chief feature which will enable you to distinguish a bug from any other type of insect is the form of the head with its piercing and sucking mouth-parts (Fig. 96). The head is prolonged at the front end into a kind of proboscis or rostrum (the other name for the order of bugs is *Rhynchota* from the Greek word ' *rhynchos*,' meaning a beak or snout). This proboscis may form an extension to the head which is bent downwards about the middle of its length, or it may be tucked underneath the head and be only visible from the side or when the insect is placed on its back (Fig. 98B). The snout is formed from the ordinary kind of insect mouth-parts in the following way ; the third pair of jaws is elongated and made into a jointed tube which is open dorsally and at its distal end. The first two pairs of jaws are long thin stylets which lie inside the tube made by the third pair ; they are longer than the third pair and can extend during feeding. The second pair of these stylets are placed close together forming two long narrow channels, one leading to the alimentary canal ; there is thus no real mouth opening unless you consider the distal end of this channel as the mouth. In feeding the two pairs of stylets pierce the skin of plants or animals, and the juices are then sucked up the long thin channel by the action of a simple suction muscular apparatus in the head. Bugs are therefore only adapted to feed on liquid food ; you will soon notice in your aquarium that if a bug like *Notonecta* kills and eats one of your more precious insects it does not eat the outer part of its victim, but merely sucks all the juices from the inside.

The water-bugs live on or in water all their lives ; they are not just aquatic for part of the time like dragon-flies, may-flies, etc. The adults may have two pairs of wings, the first pair of which are stiffened slightly to form a protective covering for the second larger membranous pair. Many of the pond skaters are, however, wingless in the adult stage. The nymphs are very like the adult into which they will grow, in fact the only difference is one of size, wings, sometimes colour, and the inability of the nymphs to reproduce. The eggs may be laid inside plant tissues, singly on submerged plants, or in gelatinous masses attached to stones or plants. After hatching from the egg the nymph normally goes through five moults before it becomes adult. After the second moult the wings begin to show on the thorax if these are present in the fully-grown insect.

The families of water-bugs are divided into those which live under the water, the Water Boatmen (*Notonectidæ*),

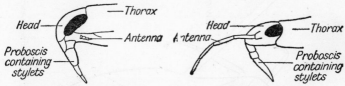

FIG. 96.—Head of *Naucoris*, side view, much enlarged.

FIG. 97.—Head of Pond Skater (*Gerris*), side view.

Two Water-Bugs showing the two types of antennæ.

Water Scorpions (*Nepidæ*), etc., and those which live always on the surface, the latter being known as pond skaters (*Gerridæ*). In the first group the antennæ are very small, and are hidden at the side of the head (Fig. 96). On the other hand, in the surface dwellers (e.g. *Gerridæ*) the antennæ are larger (Fig. 97). Beginning with the first group, the British families are the *Notonectidæ*, *Naucoridæ*, and *Corixidæ*.

Family Notonectidæ (*Waterboatmen or Backswimmers*).

This family is represented in this country by two genera, *Notonecta* and *Plea*. *Notonecta* is a very strongly made bug about 1½ cm. in length. The body is triangular in cross-section, being keeled along the back ; the colour is some shade of pale brown above, with a dark almost black under surface. The bug is a strong swimmer, most of the work being done by the third pair of legs which have long fringing hairs (see Fig. 98). This bug and members of the genus *Plea* always swim on their backs which is an easy way of distinguishing them from other bugs on sight. *Notonecta* feeds on animals often much larger than itself, such as frog tadpoles and small fish, but it will

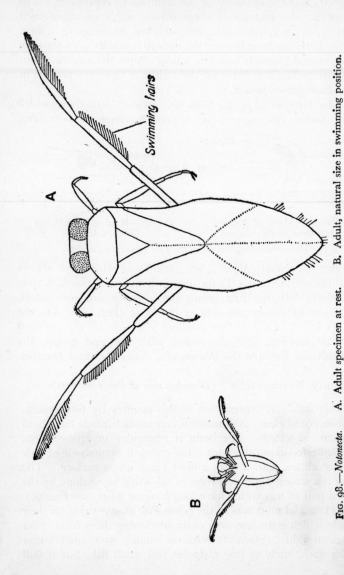

Swimming hairs

Fig. 98.—*Notonecta.* A. Adult specimen at rest. B. Adult, natural size in swimming position.

also eat other insects, and it must on no account be placed in an aquarium with any animals which you wish to remain alive. If you handle *Notonecta* you should do so carefully because it can give you quite a painful prick with its stylets. *Notonecta* is very common in ponds and near the edges of larger areas of water,—with one sweep of a net it is often possible to capture several specimens. They come to the surface periodically to renew their supply of air which they carry between a series of hairs on either side of the abdomen. They rest under water with the dorsal side normally uppermost, the head end at a lower level than the tip of the abdomen, and with the third pair of legs held straight forwards as in Fig. 98A. The front pair of legs hold on to some submerged object to prevent the animal from floating up to the surface. The eggs are laid singly on water-plants, three-quarters of the egg being inside the plant tissues. The newly hatched nymph escapes through the free end. There are six British species.

There is only one species of *Plea* in Britain, *P. leachi* (Fig. 99). This is a small bug only about 3 mm. long with a broad somewhat flattened body. It is found in large numbers together among vegetation in still water. It feeds on small crustaceans such as water-fleas and *Cyclops* (see p. 97). The oval eggs are laid singly in the tissues of water-plants.

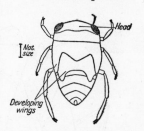

FIG. 99.—Full-grown nymph. of *Plea leachi*. After Rousseau.

Family Naucoridæ

The genera *Ilyocaris* (*Naucoris*) and *Aphelochirus* are each represented by one British species. *Ilyocoris* (*Naucoris*) is a largish bug being about 1½ cm. long. The body is oval, rather concave above and rounded below (see Fig. 100). It is found in numbers together in almost any type of still or slow-flowing fresh water. The eggs are laid partly enveloped by plant tissues. *Aphelochirus æstivalis* appears to be local, its body is elliptical in outline and much flattened, the adults usually have no wings. This bug lives in swift rivers and is found under stones in deep water or on the underside of floating leaves. The eggs are laid singly or in groups on stones, or the shells of molluscs. One mollusc, *Bythinia tentaculata* (p. 233), retaliates by laying its eggs on *Aphelochirus!*

Fig. 100.
Naucoris full-grown nymph.
After Rousseau.

Family Nepidæ.   *The Water Scorpions*

There are two genera, *Nepa* and *Ranatra*, members of which are very common in Britain.   Both   are   characterised   by possessing a long breathing tube at the hind end of the body.   *Nepa cinerea*, the only European species, is large, adult specimens reaching a length of $3\frac{1}{2}$ cm. It is a dark brown flat bug resembling a

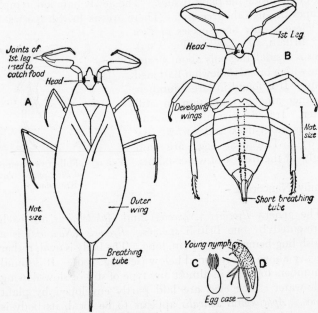

Fig. 101.—The Water Scorpion (*Nepa cinerea*).
A. Adult.   B. Nymph.   C. Egg.   D. Young hatching.

dead leaf (see Fig. 101).   It moves very slowly except sometimes when it is removed from the water, it then either pretends to be dead, or crawls away quite rapidly. It is found in still water among aquatic plants, and it feeds on other insects and any animals of a suitable size. The prey is seized by the front pair of legs which are very curiously constructed, and are held in a suitable attitude to seize the victims and hold them in a vice (see Fig. 101A, B). The breathing tube is periodically placed so that its tip is out of the water, but this is apparently not the only method of getting a supply of oxygen because the adults live quite well if the tube is removed. The eggs are laid on water-plants with seven or nine long appendages attached to one end of each. The nymphs are found throughout the summer and they seem to prefer to climb about on floating vegetation. The breathing tube of the nymph is very short and the body in the earlier stages is hairy. Specimens of *Nepa* are quite often found with small bright red oval bodies attached to them. These are a parasitic stage in the life-history of a water-mite (see p. 221).

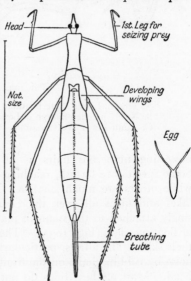

The other genus *Ranatra* (sometimes called the water-stick insect) is not so common as *Nepa*. The body is long and thin with slender legs, the first pair of which are also used in catching food. The length of an adult is about 5 cm. It is found among vegetation in still water and is difficult to see when surrounded by plant débris. There is only one European species *R. linearis* (Fig. 102),

Fig. 102.—The Water Scorpion (*Ranatra linearis*), last stage nymph.

which in most of its habits resembles *Nepa* very closely. The eggs are laid on floating plants, and each has two appendages.

Family Corixidæ.   *The Lesser Waterboatmen*

This family is represented by five genera, in Britain, which have between them some thirty-three different species.   These bugs are somewhat similar to Notonecta, but they do not swim on their backs and the dorsal surface is flat.   They normally spend most of their time near the bottom of ponds holding on to plants by their second pair of legs, but they have to swim to the surface to renew their supply of air which is stored in a depression on the ventral side of the abdomen.   They are strong swimmers, the third pair of legs

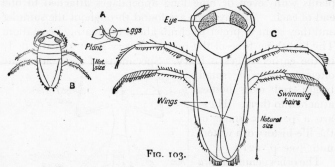

FIG. 103.

A. Eggs of Lesser Waterboatman (*Corixa*).
B. Newly hatched nymph of Lesser Waterboatman (*Corixa*).
C. Adult Lesser Waterboatman (*Corixa*).

possessing numerous hairs as in *Notonecta* (see Fig. 103). Some species are largish, reaching a length of about 1½ cm., while others are much smaller.   They are found in ponds, lakes and tarns all over the country.   The eggs are laid singly attached to the stems of plants or fixed to the threads of floating algæ (Fig. 103A).   The nymphs breathe through their skin.   The adults, particularly the males, make a shrill noise by rubbing hairs on the front legs against ridges on the head.   Most of these bugs do not pierce with their mouthparts to obtain food, they suck up particles of debris using their short proboscis like a vacuum cleaner.

To identify species see

A Key to the British Species of Corixidæ, by T. T. Macan (1939).   Freshwater Biological Association of the British Empire.   Scientific Publication No. 1.

Family Hydrometridæ, Gerridæ, Veliidæ, Mesoveliidæ.
The Pond Skaters.

All the members of this family live on the surface of the water,
and most of them can make sudden move-
ments by means of which they skim along
the water which is why they are called
" skaters." Their undersides are covered
with a dense pile of hair which prevents
them from being wetted ; some of them
enter the water occasionally. Their antennæ are very distinct and are composed
of four joints.

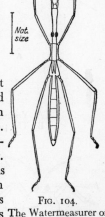

Nat.
size

The Watermeasurer or Water-gnat
(*Hydrometra stagnorum*) is a very attenuated
pond skater, about 1 cm. long, found on
the surface of stagnant water (see Fig. 104).
it walks slowly about on the surface without apparently ever entering the water.
The head of this insect is very curious
because it is much elongated. It feeds on
small insects such as flies and crustaceans
like water-fleas. The female lays her eggs
individually on plants above the water-
level, they are cylindrical, about a millimetre long, and are fixed by a short pedicil.

Fig. 104.
The Watermeasurer or
Water -gnat
(*Hydrometra stagnorum*).
After Imms.

Most adults have no wings, though a rare winged form exists.

The Pond Skater (*Gerris*) is very commonly found in large
numbers sliding along or jumping on the surface of ponds
and ditches. It has a thin flattish body with relatively
very long legs. The length of the body is from 0.65 to 1.7
cm. (see Fig. 105). All stages from young nymphs to adults
are found in summer, but it is rather difficult to tell when you
have got an adult because some of them are wingless, some
have very short non-functional wings, some have almost
fully developed wings (see Fig. 105B), and a few have wings of
full length. This means that a large specimen with half-
grown wings may be an old nymph or it may be an adult.
They feed on dead bodies of insects which land on the water
surface. The eggs are laid in groups covered by mucilage

and attached to submerged plants. There are seven British species.

The water cricket (*Velia currens*) is smaller than the other

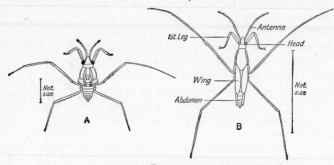

FIG. 105.
A. Young nymph of Pond Skater (*Gerris*).
B. The pond skater (*Gerris*), an adult specimen.

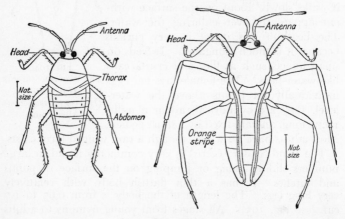

FIG. 106.
Nymph of Water Cricket (*Velia currens*).
An adult Water Cricket (*Velia currens*), wingless type.

pond skaters and its body is stouter (see Fig. 106). It is dark brown with two orange lines down the back of the adult ; both adults and nymphs have the underside of the abdomen coloured orange. It is very common on still and running water where numerous specimens are found associated together. In running water it is found near the banks and in back waters where the current is not swift. It is a good climber and will easily climb out of aquaria unless the top is covered with muslin. The wingless form of the adult is the common type, but a winged form is also found. Like *Gerris* they feed on insects or any small animals which lie on or near the surface of the water. The female lays masses of long cylindrical eggs on floating vegetation.

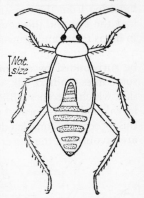

[Nat. size]

FIG. 107.—*Mesovelia.*

Microvelia is another small bug found on floating pond weeds or wet sphagnum moss. Both winged and wingless forms occur. Three species are found in the east and south of England.

Mesovelia is a much less well-known type of pond skater, single individuals of which occur among the vegetation of slow rivers, ponds, and lakes (see Fig. 107). It is greenish or blackish in colour and is difficult to see against a background of floating pond weeds. It runs along the surface and is exceedingly agile. One species, *M. furcata.*

Family Hebridæ

The insects in this family are very small ; two species, *Hebrus pusillus* and *Hebrus ruficeps*, are found on the floating pond weed *Lemna*, and on wet sphagnum moss.

To identify genera of Nymphs see

Les larves et nymphes aquatiques des insectes d'Europe, by E. Rousseau. Vol. I, Bruxelles, 1921.

For further reference

A Biology of the British Hemiptera-Heteroptera, by E. A. Butler. London, 1923.

A Key to the British Water Bugs (Hemiptera-Heteroptera excluding Corixidæ), by T. T. Macan. Freshwater Biological Association of the British Empire. Scientific Publication No. 4 (1941).

The Alder-flies (Order *Neuroptera*).—We now come to the second division of the insects which comprises those orders which have a more complicated life-history. The young on hatching from the egg are often very unlike the adult which laid the egg and this young stage is called a *larva*; its purpose in life is to eat and grow as rapidly as it can. When full grown a larva enters upon a resting stage called the *pupa* which may be surrounded by a protective covering or cocoon, which is sometimes made of silk, while in others it is just the last larval skin which becomes stiffened. During the period of rest the larval body is re-formed into that of the adult insect. When this process is completed and conditions such as temperature are favourable, the insect emerges from its pupa as an adult.

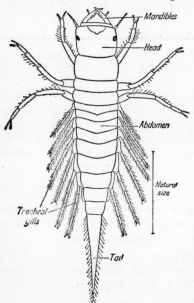

FIG. 108.—Larva of the Alder-Fly (*Sialis*).

A short time is usually necessary for the wings to expand and dry before the insect is able to fly away. In many cases the winged adults feed very little, they live on the food reserves stored up from their larval life, and their main functions are to reproduce the species and to distribute it some distance from the pond or river where they themselves were reared.

The larvæ of alder-flies are among the commonest animals living in ponds and streams. They have stout brownish bodies showing typical insect characters (see Fig. 108), but in addition to three pairs of legs on the thorax each abdominal segment possesses a pair of long jointed processes which look very much like legs. These are the tracheal gills. The abdomen is prolonged into a tapering tail-piece. It is possible at first sight to confuse an alder-fly larva with that of a beetle, especially the larva of a whirligig beetle (see p. 166). Young alder-flies are found in ponds and very slow-flowing water where the bottom is covered with mud and they usually prefer some vegetation. Small larvæ are much more active than full-grown ones and they can swim vigorously. They are carnivorous, eating many other animals among which caddis larvæ (see p. 192) and may-fly nymphs seem to be favoured. After about a year or perhaps longer in the water, during which the larva has moulted a number of times, it reaches its full size. It then crawls out of the water for a distance of several yards until it finds a suitable piece of earth in which to dig a small cavity. It crawls inside this, sheds its last larval skin, and pupates. In two or three weeks' time the adult fly emerges and it is found very commonly flying rather clumsily near ponds and

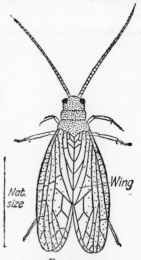

Fig. 109.
Adult Alder-Fly (*Sialis*).

rivers in May and June. The adult (Fig. 109) has large coarse wings with clear black veins, and long jointed antennæ. It is not a strong flier and is often found resting on walls, tree trunks, and plants near water ; at rest the wings are folded in a characteristic way forming a kind of roof across its back. The female lays very many eggs in batches, sometimes 2,000, on plants near water. When the young hatch out they have to find the water or perish. There are two British species, *Sialis lutaria* the common one, and *Sialis fuliginosa*.

The Spongilla Fly.—The Spongilla fly (*Sisyra*) is really a lacewing fly related to the kind with pale green wings and bronze eyes which often fly into houses in summer. Its larval stages are spent in water as a parasite on fresh-water sponges. It is found in still water and in streams clinging to the surface of and embedded in the sponge from which it extracts its food. It is not found commonly, but this may be partly due to its being small and inconspicuous. It is only about half a centimetre long when full grown, and its soft hairy body is the same colour as the sponge, dirty white or greenish (see Fig. 110). On the underside of the larva are seven pairs of finger-like gills, one on each of the first seven abdominal segments. The larva leaves the water to pupate and it spins a cocoon of silk attached to some support ; inside this it changes

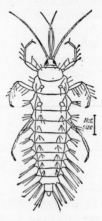

FIG. 110.—Larva of Spongilla Fly (*Sisyra*). After Rousseau.

into a pupa. After the adults have emerged the females lay their eggs in small clusters on any surface which is near or overhangs the water. There are three British species of which *Sisyra fuscata* is the commonest.

There is one other genus closely related to the Spongilla fly whose larvæ are found in damp moss at the edge of streams, or actually in the water, this is the genus *Osmylus*. It is rather like *Sisyra* in appearance, but much bigger, $1\frac{1}{2}$ cm. long when full grown with a more narrow body and no gills. The only species, *Osmylus chrysops*, feeds on Chironomus larvæ (p. 210).

To identify British species of Alder-flies

Keys to the British Species of Aquatic Megaloptera and Neuroptera, by D. E. Kimmins. Pub. (1944) Freshwater Biological Association of the British Empire. Scientific Publication No. 8.

General. Biologie der Süsswasserinsekten, von Dr. C. Wesenberg-Lund (1943). Berlin, Wien.

CHAPTER 9

THE INSECTS (*continued*)

The Fresh-water Beetles (*Coleoptera*).—The beetles are an enormous group of insects containing a few families which are entirely aquatic ; there are also some aquatic genera among families of mainly land-dwelling species. Some of these latter only spend part of their lives in water. All beetles have a complex life-history, involving the stages of egg, larva (which is often rather grub-like), pupa, and finally the adult winged insect.

In their adult stage beetles are not difficult to recognise ; some well-known, non-aquatic examples are cockchafer beetles, dor beetles, and ladybirds. The body is usually hard and often shiny, the head shows clearly the typical insect characters mentioned on page 115 ; antennæ, eyes, and mouth-parts are usually all well developed ; the jaws are adapted for biting. Sometimes the first pair of jaws (the mandibles) are very large, each being a conspicuous sickle-shaped appendage. The first segment of the thorax is normally very large while the next two segments are hidden by the wings when viewed from above. The first pair of wings is specially modified in beetles to form a horny protective covering for the hind pair, which are large and membranous. This is a character which never fails to identify any insect possessing it, as a beetle. This first pair of horny wings may be as long as the body or considerably shorter ; they fit down tightly against the back of the insect, and it is only when they are carefully lifted up by some instrument that you can see the much larger membranous pair of wings folded up underneath. All the flying is done by this second pair, and if you watch a beetle, such as a ladybird, alight on some object after flying, you can see the second pair of wings being folded up under the first pair. The legs of land beetles are usually thin, hard, and very clearly jointed ; they are in fact a walking or running type of leg. Among the adults of aquatic beetles the legs are often somewhat flattened, especially the last pair, and their edges are fringed with long hairs which help in swimming. The abdominal part of the body may be completely hidden above by the wings ; it probably consists of ten segments but only between five and eight are clearly visible, the others being either fused or telescoped together.

The larvæ of beetles are variable in form, and for that reason may be difficult to recognise. A glance at the drawings on the following pages will give you some idea of the various kinds. Larvæ which are active swimmers have relatively hard bodies with prominent heads and well-developed legs (such as Dytiscus and whirligig beetle larvæ, Figs. 111, 121). Less active kinds tend to have soft bodies with well-developed heads, but very small legs (such as the larvæ of the silver water-beetles). All have biting mouth-parts with, in the case of carnivorous forms, one pair of jaws (the mandibles, see Figs. 71, 127) very large. The antennæ are generally quite large, in one family (the Helodidæ) they are very long indeed. The abdomen bears a number of side processes or gills in a few cases, and the last abdominal segment may have two " tails " or cerci. There are ten abdominal segments. The aquatic larvæ may breathe dissolved oxygen in the water through their gills, or they may come to the surface to obtain oxygen from the air. Some members of one family obtain their supply from the air enclosed in submerged plant tissues. The resting pupal stage is normally passed through on land near water. Often the full-grown larva constructs an underground cell in which to pupate, so that the pupa is a much less familiar object than the larva or adult.

Water-beetles are either plant feeders or are carnivorous.

There are four British families, of which all the species are aquatic ; in addition a few members of four other families also live in water.

Completely Aquatic Families

Family Dytiscidæ

Belonging to this family there are about 106 British species. The larvæ and adults are found in all types of water, but pupation takes place in earth near the water. The adults are all good fliers so that dispersal is easily accomplished by this method. They may be distinguished from other families by their thread-like antennæ and strong mandibles, all Dytiscids being carnivorous both as adults and larvæ. The hind legs of the adults are flattened and are fringed with long swimming hairs. The front legs of the males of some species have three of the terminal joints made into a kind of sucker (Fig. 111C) with which they can hold on to the females while mating. The horny wings are as long as the abdomen, and in the females they often show longitudinal ridges. The adults breathe by coming to the surface and placing the extreme tip of their abdomens out of the water. At this end of the body there is a large pair of spiracles (breathing holes) which come in contact with the air. A supply of air is also stored in a space between the horny wings and the body ; this is used while the beetle is submerged and is renewed each time it comes to the surface.

The larvæ of the Dytiscids are all of the same type though some are more active than others, and the length of the two tail processes (cerci) varies (Figs. 114, 115). It is not possible here to give a description of the larvæ which will enable you to identify the genus of a specimen found, since the larvæ of British species have not all been accurately described, but they all have a well-chitinised head with sickle-shaped mandibles, two claws at the end of each leg, these last appearing " feathery " due to the presence of swimming hairs.

The large family of *Dytiscidæ* is represented in Britain by eighteen different genera of which the common ones are described below.

Genus Dytiscus.—There are six British species, of which *Dytiscus marginalis* (Fig. 111) is much the best known. This beetle, sometimes called the Carnivorous Water-beetle or the Great Diving beetle, is very large, olive-brown in colour, and is found commonly in all kinds of ponds and small reservoirs. The adults will eat tadpoles, snails, insects, and even each other ; they live very well in captivity, and will not attack one another if they are not overcrowded and if they have plenty of other live creatures to eat. The male is easily distinguished from the female on account of his having a round sucker-like disc on the first pair of legs. The hind pair of legs of both sexes has long hairs which aid in swimming. It is well worth while watching these animals' swimming movements in a glass tank, and also the way in which they

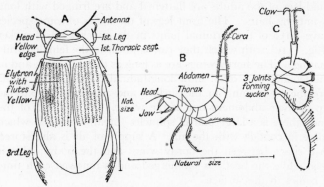

Fig. 111.—The Great Diving Beetle (*Dytiscus marginalis*).

A. Adult female.
B. Larva in characteristic position.
C. Tarsal joints of male, first leg much enlarged.

come to the surface every few minutes to get a fresh supply of air. The time that you are most likely to catch this beetle in a pond is when it comes up to the surface for breathing. The female lays her eggs singly inside the stems of submerged water-plants in spring. The larva (see Fig. 111B) is as ferociously carnivorous as is the adult, and when it is approaching its full size it will also attack tadpoles, snails, and any other soft-bodied creatures of its own size or larger. It has very long sharp-pointed mandibles which are opened out wide when the insect is on the look-out for food. These pierce the soft tissues of the prey when they close on a victim. The actual method of eating is peculiar ; the mandibles are pierced for the whole of their length by a canal which leads from the tip of the mandible to the fore part of the gut (the pharynx). After the food has been seized, a digestive juice is passed down the two canals into the body of the prey, this partly digests the contents of the body and breaks it down into a liquid soupy mass. By means of a pumping action, which is done by the muscles of the pharynx, this liquid is pumped up the canals in the mandibles into the alimentary canal of the beetle larva. When the larva has finished its meal nothing but the skin and any hard parts of its victim are left ; these are cast aside. The larva swims with its legs, which are hairy, but it can also make more violent movements by jerking its abdomen. The larva is lighter than water so that it automatically floats up to the surface if it is not swimming or holding on to some submerged object. The tail cerci and last two abdominal segments are also hairy, which allows the larva to hang head downwards from the surface film of the water in such a way that the tip of the abdomen, on which are two spiracles, is in contact with the air which the insect breathes. Leading away from the two tail spiracles are two long breathing tubes or tracheæ which traverse the length of the body giving off branches to supply all the tissues. As *D. marginalis* is much the commonest large dytiscid beetle, if you find a larva measuring 4-5 cm. you may be reasonably sure that it is that of *D. marginalis*, but of course the larva is not that size when it first hatches. A fully-grown larva leaves the water and hollows out a small cell in some damp earth nearby. Inside this it moults and changes to the resting *pupa* stage, in which state it may remain for a few weeks, or through the whole winter, depending on the temperature at the time of pupation in late summer. The adult when it emerges finds its way to the water, but it is a good flier and does not need to spend the rest of its life of several years in the same pond. The adults hibernate during the winter in the mud at the bottom of ponds ; they appear to be dead during hibernation, but soon wake up if placed in a warm room.

Another species of *Dytiscus* which has received some study is that of *D. lapponicus*, which has only been found in the inner

Hebridean islands of Skye and Eigg. The remaining four species are all rare and local in occurrence.

Genus Deronectes.—There are five British species of this genus of which *D. elegans* is shown on Fig. 112. This is a small brown and yellow beetle which is common in Yorkshire and probably also in many other places. The larva is often found in stagnant ponds among masses of thread algæ, and its principal diet would seem to consist of the water-louse

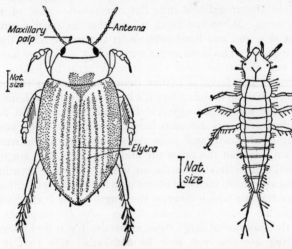

Maxillary palp

Antenna

Nat. size

Elytra

Nat. size

Fig. 112.—*Deronectes* (*Hydroporus*) *elegans*, adult.

Fig. 113.—Larva of the *Dytiscid* beetle (*Hydroporus*). After Brocher.

(*Asellus*, see p. 105) and other insects such as may-fly nymphs (see p. 133).

Genus Hydroporus.—This genus is very closely allied to the previous one, in fact they are often united together in the *Genus Hydroporus*. All the species are small, measuring only a few millimetres long, but they are common in ponds. The anterior part of the head in the larva is prolonged into a short beak (see Fig. 113). There are thirty-seven species.

Genus Hyphydrus.—There is only one British species, *H. ovatus*, but it is very common (Fig. 114). The adult measures about half a centimetre in length, and the shape of the body being practically globular makes this beetle easy to recognise ; the colour is reddish. The larva has a very pronounced snout at the head end (Fig. 114).

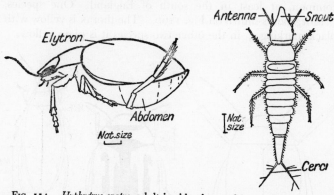

FIG. 114.—*Hyphydrus ovatus*, adult in side view and larva, dorsal view.

Genus Acilius.—One of the two species of this genus, *A. sulcatus*, is common in ponds (Fig. 115). The adult is very like a *Dytiscus*. It is about 1½ cm. long, with thread-like antennæ and a flat body which has transverse black markings on the thorax. The front legs of the male bear suckers ; the third legs are much clothed with swimming hairs in both sexes. The larva has a specially long first thoracic segment (see Fig. 115).

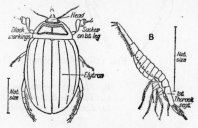

FIG. 115.—The furrowed *Acilius* (*Acilius sulcata*).

A. Adult. After Joy.
B. Larva. After Brocher.

Genus Ilybius.—There are seven species of this genus two of them fairly common. The outline of the body is longer and narrower than in the other Dytiscids with the tip of the abdomen showing beyond the elytra. The Mud Dweller (*I. ater*) is a common southern species with a narrow black body 1 cm. long. Near the sides of the elytra are two longitudinal yellow streaks (Fig. 116c).

Genus Laccophilus.—Beetles of this genus are probably fairly common, at least in the south of England. One species, *L. variegatus*, is shown on Fig. 116D. The thorax is yellow with black markings ; in the other two species it is plain yellow.

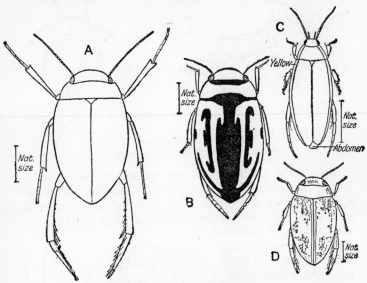

FIG. 116.—Four common Dytiscid Beetles.

    A. *Agabus.*
    B. *Platambus maculatus.* After Joy.
    C. The Mud Dweller (*Ilybius*).
    D. *Laccophilus variegatus.*

Genus Agabus.—To this belong nineteen species, some of which are very common in ponds. The average size is rather under 1 cm. long ; the body is very shiny, and of a dark reddish or blackish colour (Fig. 116A).

Genus Platambus.—The only species, *P. maculatus*, is fairly common. The adult is reddish-yellow, with curious black markings on the body (Fig. 116B). Length is about 8 mm. long and it is generally found in streams.

Family Hygrobiidæ. *The Screech Beetle*

In Britain there is only one species of this family, known as the Screech Beetle (*Hygrobia tarda*, late *Pelobius*). It is found

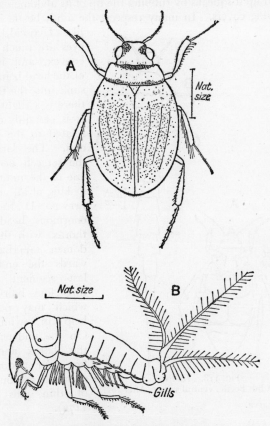

Fig. 117.—The Screech Beetle (*Hygrobia (Pelobius) tarda*).
A. Adult. After Joy.
B. Larva. After Schiödte.

in ponds over England as far north as Yorkshire and in some parts of Ireland. The adult has a very strongly convex body about a centimetre long. The colour is reddish-brown with the elytra (wing covers) blackish. The front and hind edges of the thorax are black. The legs have swimming hairs but are otherwise unmodified (Fig. 117A). When this beetle is picked up it squeaks by rubbing the tip of its abdomen against the wing covers. In many respects the adult beetle is very

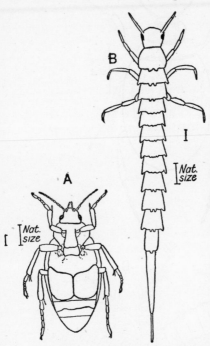

like a Dytiscid, but the eyes are much more convex, and instead of the head being well sunk into the thorax, there is a slight neck. It is certainly closely related to the Dytiscids. The larva of the screech beetle is one of the most odd-looking of all aquatic larvæ. It has an enormous head and thorax, with the abdomen tapering towards the end, the last segment bears three long hairy tails which may possibly help in swimming. On the underside of the thorax and the first four abdominal segments are bunches of thin gills (Fig. 117B).

Fig. 118.—*Haliplus*.
A. Adult Beetle, ventral view. After Joy.
B. Larva.

Family Haliplidæ

This family is also closely allied to the *Dytiscidæ*. All the beetles are small, but they are easily recognised by examining the first joint of the third pair of legs in the adult; this is expanded into a large plate (Fig. 118A). There are eighteen British species, sixteen of which belong to the genus *Haliplus*, of which some are common. The adults and larvæ are chiefly found among thick vegetation of the filamentous algal type. Thick floating masses of *Spirogyra* are favourite haunts. The adults are yellowish or reddish in colour and are never more than 6 mm. long. The larvæ (Fig. 118B) are small with a long thin body ending in a single-tail process. The head has poorly developed eyes and a pair of short antennæ. The legs are small and the larva is very inactive. This larva feeds on filamentous algal plants such as *Spirogyra*. The first pair of jaws (mandibles) are pierced by a canal as in many Dytiscids ; these are inserted into a *Spirogyra* cell and the contents are sucked up the canals into the mouth. When one cell is finished the next in the filament is treated in the same way, so that eventually a whole chain of empty cells is left.

Besides the genus *Haliplus* there is one species belonging to the genus *Cnemidotus*. The larva of this beetle has a pair of tracheal gills and a pair of long processes arising from the dorsal side on each segment so that it appears to be covered in filaments. The last segment has a long pair of tails. Both this beetle and one more belonging to the same family, *Brychius elevatus*, are rare ; the latter is chiefly found in running water.

Families Helmidæ (Elmidæ) and Parnidæ

These are two small families of exclusively aquatic beetles which are sometimes united under the name of Dryopidæ, They are all vegetarians, small in size, and live in swift-flowing rivers and streams.

Family Helmidæ

The adults of this family are all only 3 mm. long, or less. They are found under stones in running water. They have retiring habits, and are only active at night when they walk about on the bottom of the stream or among the vegetation at the surface. They do not swim, but are able to grip the surface of stones by means of a very long last joint to their feet which bears a pair of long sharp claws (Fig. 119A). The larvæ are found in the same places as the adults, and the life-cycle is thought to last a year. The larvæ have hard dark brown bodies with a characteristic feathery process at the posterior end which is usually arranged in three groups (Fig. 119B). These are constantly being retracted inside the body then extended; they are tracheal gills used for breathing. In the genus *Helmis* the larva is broad and flat, while in the other genera of *Macronychus*, *Riolus*, *Latelmis*, *Esolus*, and *Limnius* the larva is narrow and worm-like (Fig. 119C). Most genera are local or rare, but *Helmis* is common in Hertfordshire streams.

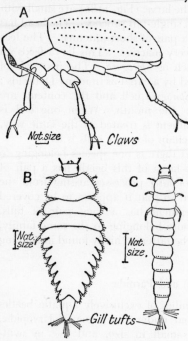

FIG. 119.—Family *Helmidæ*.

A. Adult *Helmis*.
B. Larva of *Helmis*.
C. Type of larva found in genera other than *Helmis*. All after Brocher.

Family Parnidæ

This is represented in Britain by seven species of *Dryops* (*Parnus*), two of which are fairly common, and by one rare species of *Helicus*. The adults are small, only 4-5 mm. long, with hairy bodies. They can be recognised by their curious antennæ (Fig. 120A). They come out of the water and fly about at night; during the daytime they are found in streams. The eggs are laid on plants out of water on the banks of the stream; the worm-like larvæ are aquatic, they have very short legs, and are lethargic in habits. The last segment is flattened with a valve-like structure at the end (Fig. 120B).

Family Gyrinidæ. (*The Whirligig Beetles*)

In this family the species are all much modified for living in water. There are ten species of the genus *Gyrinus* of which the common Whirligig beetle (*G. natator*) is much the most familiar. The adults of all the species are small shiny, black, oval beetles with yellow legs which congregate together and swim round in circles on the surface of slow-running streams or stagnant water (Fig. 121). If they are disturbed they dive underneath carrying with them an air-bubble. They fly readily from one pond to another, so that in captivity they must be covered up. If a specimen is examined it will be found to have short clubbed antennæ, eyes divided into two portions, and most curiously

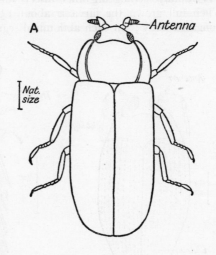
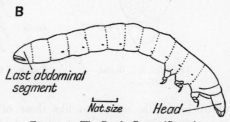

Fig. 120.—The Beetle *Dryops* (*Parnus*).
A. Adult. After Joy.
B. Larva. After Brocher.

shaped second and third pairs of legs. Each segment of these legs is greatly flattened and fringed with hairs. They are used in swimming as well as for making the characteristic skating movements on the water surface. The abdomen extends beyond the wing covers. Late summer is the time when huge numbers of adults are found together gyrating over the surface of water. The females lay their eggs in spring on the leaves of submerged water-plants. These hatch into a most peculiar larva (Fig. 121) which grows very rapidly feeding on other insects and sometimes plants. When full grown the larva is about 1½ cm. long with an elongated body ; on each abdominal segment there is a pair

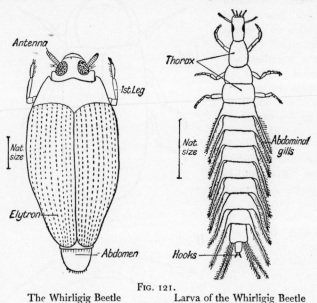

FIG. 121.

The Whirligig Beetle   Larva of the Whirligig Beetle
(*Gyrinus*).   (*Gyrinus*).

of tracheal gills, somewhat like those of an alder-fly ; the last segment has two pairs of gills and also two pairs of re-curved hooks which are said to be used for climbing. When first found the *Gyrinus* larva may be mistaken for a young alder-fly larva on account of the gills, or it is possible to con-fuse it with a caddis larva which has come out of its case, as many of these have gills and all have hooks on the last segment. About the end of July the larvæ leave the water by climbing up water-plants until they are above the water surface. They then spin a cocoon inside which they pupate. These cocoons are very well hidden and are seldom found. A month later the adults emerge and this is the time of year when they are so common. When the cold weather comes the adults hibernate in the mud around the roots of water-plants at the bottom of streams and ponds, remaining there for the whole winter. In spring they reappear on the surface and the females begin to lay eggs. The adults seem to be more common than the larvæ, though of course this is not really the case, the reverse being true. The larvæ have to be searched for with a net at the right time of year, whereas the adults are very obvious creatures on the water surface for a large part of the year.

There is one species of Whirligig which belongs to a different genus from *Gyrinus*. This beetle is very similar to the others except that the upper surface is clothed with short thick hairs. It is the Hairy Whirligig (*Orectochilus villosus*).

Families with some Genera Aquatic for all or part of their lives

Family Helodidæ

In this family of beetles all the adults live on land, but they have aquatic larvæ, and it is with these only that we are concerned. The larvæ all have flattened bodies with long antennæ.

Genus Cyphon.—The larvæ of this genus, of which there are five or six British species, are found among floating vegetation on stagnant water. The species *C. variabilis* is the commonest. The length of the body is about 8 mm., much bigger than that of the beetle into which it develops which is only 3.5 mm. long. The colour of the larva is dark brown. It does not swim, but walks about among the floating vegetation ; it breathes air through a pair of spiracles which open on the last segment of the body. This segment is placed out of the water to obtain a new supply of air ; when submerged it is covered by an air-bubble. If these larvæ are kept in an aquarium they must be supplied with floating plants as they are otherwise unable to support their weight at the water surface to obtain their air supply. On the last segment of the body there are three white finger-like processes which project from the anus, these have no tracheæ but are said to be filled with blood. They can be retracted at will. About the beginning of July the larvæ are full grown; they then come out of the water and pupate inside a cocoon among the plants round the pond. In captivity it is fairly easy to obtain the pupæ by providing the larvæ with plants which stick up out of the water. The cocoons are then formed on these.

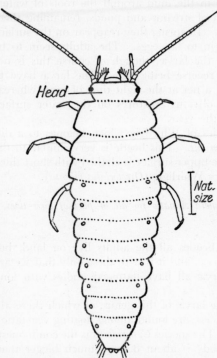

Genus Helodes.— The larvæ of this genus have the same general appearance as that of the preceding one. The body is flatter, broader at the anterior end and tapering behind (Fig. 122). The last abdominal segment has two

Fig. 122.—*Helodes marginata*, larva.

spiracular openings, and these larvæ have to come to the surface to renew their air-supply. The last segment bears five white retractile " gills " containing tracheæ which are protruded from the anus. There are two British species, *H. minuta* which is the smaller measuring about 6 mm. when full grown, and *H. marginata* which is about a centimetre long. Both have dark brown bodies, but that of *H. minuta* is much broader in proportion to its length. They are found on the undersides of stones in small rivers and mountain streams all the year round, but they appear to be commonest in winter. The small species is common all over Britain while the bigger one is more local, but where it does occur it is in abundance. There is one valley near Sheffield where I have found it in huge numbers living among tufts of leafy liverworts which are growing in the path of a small waterfall. They live well in captivity, but have the habit of escaping from uncovered dishes.

Genus Hydrocyphon.—One species of this genus is found in this country, *H. deflexicollis*. The larvæ are found under stones during summer in cold mountain streams. This is a local species, but the larvæ usually live in small communities of about fifteen individuals. In appearance the larvæ are very similar to that of *Helodes* or *Cyphon*, but they are much smaller, being only 3·5 mm. when full grown. There are five anal gills as in *Helodes*. The larva pupates under water, being attached to a stone by a silk thread. The pupa is surrounded by an air bubble which clings to the surface of the adult when it emerges so that it can reach the surface of the water unwetted. The larvæ can leave the water for short periods and return to it later. The adults live on sallows.

Genus Prionocyphon.—One very local species, *P. serricornis*, the adults of which live on beech trees while the larvæ live in water-holes on the roots.

Genus Scrites.—Two rare and local species, the adults of which are found on rushes and *Persicaria* plants. The aquatic larvæ are remarkable for their long antennæ which are half as long as the body ; their body length is about 4 mm.

Family Chrysomelidæ

This is another large family of beetles of which only a few are aquatic or semi-aquatic.

Genus Donacia.—There are nineteen British species of this genus, some of them common. The adults are terrestrial, but the females may often be found on the floating leaves of water-plants such as the white water-lily, the arrow-head, and marsh marigold, because she lays her eggs on plants such as these. She bites a round hole in the leaf surface, then passes her eggs through to the underside where they are stuck to the leaf round the edge of the hole. As soon as the larvæ hatch they sink through the water to the roots of the plants to which they become attached. The body of the larva is white and grub-like, it is a very inactive creature living on the tissues of the root, but always having the last abdominal segment penetrating by means of hollow spines which communicate with tracheæ through the outer tissues of the root into the air spaces within (Fig. 123B). (It is characteristic of water-plants to have many air spaces in the stems and roots.) From the air spaces in the plant the larva obtains its supply of oxygen. The insect pupates in a silken cocoon attached to the root surface in such a way that the root forms one side of the cocoon. A number of fine holes in this part of the root surface supply the space inside the cocoon with air, so that the pupa is able to breathe. The adult emerges from the cocoon in spring, but apparently it usually comes out of the pupal skin several months before but remains immobile inside the cocoon until the weather is suitable. As the adult ascends through the water it is enveloped by air, so that on reaching the surface it is unwetted and is able to fly off immediately. It is not usually possible to keep the larvæ in captivity, but if water-plants are pulled up by their roots in winter the cocoons may be found; if these contain the resting beetles, and not pupæ, then they will emerge in a warm room if the roots are placed in water. The adults vary in size from 7 to 14 mm. in length, all having

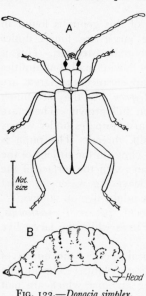

Nat. size

FIG. 123.—*Donacia simplex.*
After Joy.

A. Adult.
B. Maggot-like larva.

a strong metallic sheen to their bodies which are often some shade of blue or green (Fig. 123A).

Genus Hæmonia (Macroplea).—There are two species, one, *H. mutica*, is local while the other, *H. appendiculata*, is very rare. The adults are found under water clinging to the roots of Potomageton. They are very immobile and will stay in the same position for days. Stagnant or slow-running water is preferred, and the only way to collect the adults is to pull up the plant by the roots and examine them for specimens. The adults are to be found from August to

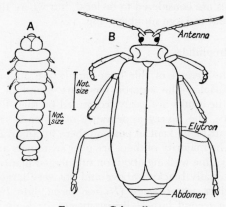

Fig. 124.—*Galerucella.*

A. Young larva.
B. Adult.

October ; they are often in pairs, the female being slightly larger than the male ; length of body is about 4·5 to 6 mm. The larva are similar to those of *Donacia*.

Genus Plateumaris.—These beetles are very closely allied to *Donacia* and they have similar habits. They are found chiefly on the leaves and stems of water-plants, but only one of the four species, *P. sericea*, is common. They may be distinguished from *Donacia* in the adult stage by their having a more convex body which is rounded posteriorly, and by their broader tibial joint to their legs. *P. sericea* is black and about 9 mm. long.

Genus Galerucella.—Very little seems to have been written about the biology of this genus. Some of the species at least are amphibious, but there seems to be some doubt as to what are distinct species and what are varieties of one species. In July and August an orange-yellow species is found on the floating leaves of the water-lily family and also on those of the Water Persicaria. Large yellow eggs are laid by these beetles on the undersides of the leaves. From these hatch jet-black larvæ which eat the leaves on which they were laid. As the larvæ grow they become banded with brown and black. These beetles are considered to be local, but where they occur they are found in large numbers (Fig. 124).

Family Hydrophilidæ

Not all the genera of this large family are aquatic, some being terrestrial. The aquatic types are found in stagnant water, and the adults are easily recognised by their flat silvery under-surface, the silvery appearance of which is caused by a supply of air being kept in position there by an arrangement of hairs. The majority of them are poor swimmers and spend most of their time walking about on submerged plants. Their bodies are usually blackish above and not very large. However, the silver water-beetle (*Hydrous piceus*, late *Hydrophilus piceus*) is the second largest British beetle, its body reaching a length of nearly 4 cm., while *Hydrophilus caraboides* is about half that size. The others are all less than a centimetre long. Practically all of them are vegetarians as adults. The females of this family construct elaborate floating egg cocoons made of silk which is produced in glands opening on the last abdominal segment, inside which they lay about fifty eggs. The larvæ, which are to be found in ponds covered with a mass of plant growth, such as duckweed, are carnivorous, with a huge pair of very conspicuous mandibles (Fig. 127). All the larvæ are very similar in appearance and habits, but their full-grown size varies of course with the size of the adults into which they develop. The larvæ usually pupate in damp earth near water.

The Silver Water-beetle (*Hydrous piceus*).—Is found in the southern counties of England and as far north as Yorkshire, but it is very local. The males differ from the females in having an enlarged last joint to their front legs. The colour of both sexes is black with red antennæ. The females spin a large egg cocoon on the surface of the water and lay about 50-100 eggs inside. These hatch inside the cocoon in about seventeen days ; they crawl round for some time before finding their way out through a specially thin place. The larvæ are very carnivorous, eating principally young snails.

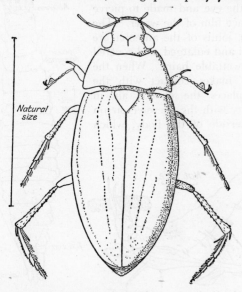

Natural size

FIG. 125.—The Silver Water-Beetle (*Hydrous piceus*). After Miall.

When they grow bigger larger snails and tadpoles form their principal diet. The larva breathes air through a pair of spiracles which open on the last abdominal segment. This segment is periodically placed above the water surface, or the larva hangs from the surface film with its spiracles in communication with the air. The full-grown larva is about 7 cm. long ; it leaves the water to pupate in damp earth beside the pond. It is only possible to get larvæ to pupate successfully in captivity if they are provided with very damp

earth.   When the adults emerge they make their way to the
water.   During the cold months they hibernate at the bottom
of the pond among the mud.   The adults renew the supply
of air, which is carried on the ventral surface of the thorax
and abdomen, in a rather curious way.   They come to the
surface periodically and place their bodies slightly on one
side.   The antenna of the side near
the surface is then bent backwards
behind the eye and made to pierce
the surface film of the water.   The
last four joints of the antenna are
flattened and enlarged, and covered
with unwettable hairs.   When the
antenna makes contact with the
air it places the outside air in
continuity with the air reservoir on
the underside of the body, which

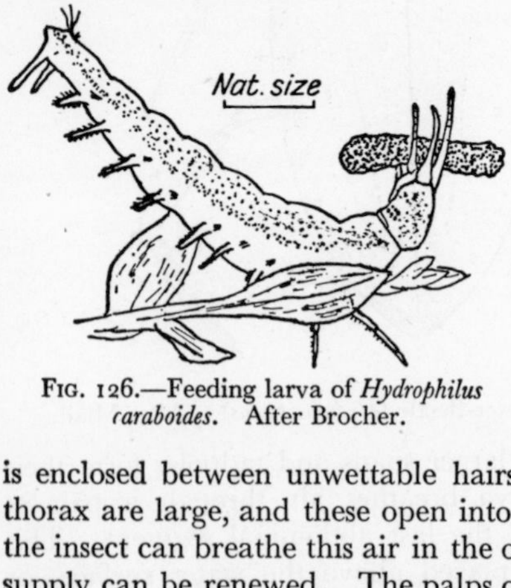

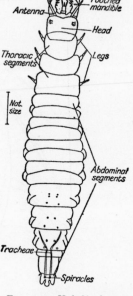

FIG. 126.—Feeding larva of *Hydrophilus*
caraboides.   After Brocher.

FIG. 127.—*Hydrobius* larva.

is enclosed between unwettable hairs.   The spiracles on the
thorax are large, and these open into the air reservoir, so that
the insect can breathe this air in the ordinary manner and the
supply can be renewed.   The palps on the second pair of jaws
(maxillæ) are greatly lengthened, and these seem to have taken
over the functions ordinarily carried out by antennæ (Fig. 125).

Genus Hydrophilus (late *Hydrochares*).—There is one species,
H. caraboides, which in appearance and habits is very similar
to the silver water-beetle except that it is much smaller, being

only 18 mm. long. It is found in the same localities as the larger species and like it is very local (Fig. 126).

Genus Hydrobius.—There is one species, *H. fuscipes*, which is common in stagnant water (Fig. 129C). It is also very like the silver water-beetle in appearance and habits but it is very much smaller, being only about 8 mm. long. Sometimes it has a strongly bluish or greenish metallic sheen. The larva (Fig. 127) is characterised by having four teeth on each of its large mandibles ; there are no lateral processes on the

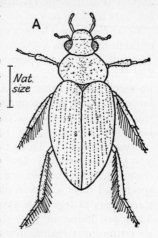

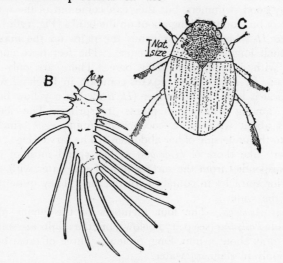

FIG. 128.—Hydrophylid Beetles.

A. Adult of *Berosus æriceps.*
B. Larva of same.
C. *Laccobius minutus.* All after Brocher.

abdomen as in some Hydrophilid larvæ, and the last segment has only a single pair of side appendages.

Genus Berosus.—There are several species of this genus, none of them more than 5·5 mm. long and all of them local or rare. The adults have conspicuous swimming hairs on the second and third pairs of legs, and they have much greater swimming powers than any other member of the family (Fig. 128A). The larva has a pair of enormously long filaments, which are really tracheal gills, on each of the abdominal segments except the last which has none (Fig. 128B).

Genus Laccobius.—These beetles are all small, varying from 3 to 5 mm. in length. Out of the seven British species two are common, while the rest are rare or very local. The common species all have a very rounded outline, and the second and third pairs of legs have swimming hairs on the tarsal joints. They are found in clear-standing water, or occasionally in slow-running streams, living on the bottom. They are reasonably good swimmers, but they can come out of the water, and may be found walking about on the banks (Fig. 128c).

Genus Helochares.—In this genus the palps on the maxillæ are much longer than the antennæ. The feet have curved claws with a tooth near their base. There are only two species : *H. lividus* (Fig. 129A) is common in England while *H. griseus* is local. The former (*H. lividus*) is a yellow beetle found among submerged plants in spring. The female carries about with her a cocoon containing eggs which is fixed to the underside of her abdomen. The larvæ (Fig. 129B) are very like those of *Hydrobius* ; they spend most of their time suspended from the surface film of the water with their posterior spiracles in contact with the air. They quite often leave the water.

Genus Anacæna.—The four members of this genus much resemble *Philydrus* (see p. 177 below), but the adults are smaller being only about 3 mm. long, or less. Three of them occur commonly in stagnant water.

Genus Helophorus.—The adult beetles of this genus are easily recognised because the first thoracic segment has clear longitudinal striations. There are about seven British species, most of them being amphibious in the larval state, but some

are terrestrial.   The adults are usually found out of the water,
sometimes they may be taken on the surface.   The largest
species is *Helophorus aquaticus* (see Fig. 129D) which is moderately
common.   In the larva the head has large mandibles and is
otherwise fairly typical for the family, but the three thoracic
segments are strongly protected by hard skin ;   there are two
tails at the end of the abdomen (Fig. 129E).

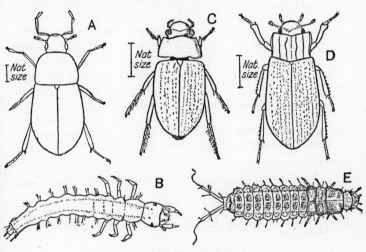

FIG. 129.—Hydrophilid Beetles.

A. *Helochares lividus.*   After Joy.
B. *H. lividus* larva.   After Brocher.
C. *Hydrobius fuscipes.*   After Miall.
D. *Helophorus aquaticus.*   After Joy.
E. *Helophorus grandis* larva.   After Brocher.

Genus Philydrus.—These beetles are also fairly small, 4-5 mm.
being the average length.   There are about seven British
species, two of them common.   The antennæ are usually
red with a black club.   The female makes an egg cocoon
shaped like a slipper or a round casket (Fig. 130) which is
attached to floating plants.   The larvæ are also much like
that of *Hydrobius*, but they have six pairs of false legs on the
abdomen.   They do not swim but crawl about among floating

12

plants, quite often coming out of the water. The larva, unlike those of other water-beetles, does not pupate in damp earth. The pupa is found hanging upside-down attached to mosses on the edge of the pond. Near to the pupa on the moss is often found the last larval skin (Fig. 131).

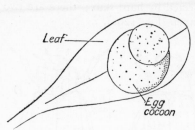

FIG. 130.—Egg mass of *Philydrus*.

Besides those Hydrophilid beetles already described there are a number of other genera, such as *Octhebius*, *Hydrochus*, and *Hydræna*, the members of which are found in or at the edge of water. All of these are very small, less than 3 mm. long, and with the exception of one species (*Hydræna riparia*), they are rare or local. Nothing is known about the larvæ of *Hydrochus* and *Hydræna*.

L'Aquarium de Chambre, by F. Brocher (Paris, 1913), gives a good synopsis of aquatic beetles.
For detailed information about adult beetles see
British Coleoptera (2 vols.), by N. Joy, London (1932)
British Water Beetles, by F. Balfour-Browne (1950). Royal Society Monograph.

Aquatic Moths (Lepidoptera).—The order of insects which contains the Moths and Butterflies is closely allied to that of the Caddis-flies. The adults of both orders are much alike, the chief similarity being their two pairs of wings which have the same type of venation and are covered with hairs and scales. The larvæ are also alike in many features, but whereas all caddis-flies have aquatic larvæ, there are only very few moths whose larvæ live in water.

Among British moths there are several small species (all belonging to the *Microlepidoptera*) which are known as China Mark Moths. These belong to the genus *Nymphula* (*Hydrocampa*), and their larvæ live on water-plants, being found in many cases under the water. These larvæ chiefly frequent the leaves of water-lilies, the Arrow-head and Potomageton. In appearance they are very like an ordinary

caterpillar with three pairs of true legs on the thorax, and
five pairs of false legs on the abdomen (Fig. 132). They hatch
out from eggs which are laid on the underside of the leaves.

In some species the young
larvæ on hatching descend
to the bottom to live on
the stems of the water-
plants while they are small,
later coming up to the top
to live among the floating
leaves. Others appear to
live among the floating
leaves from the beginning.
The larvæ do not walk
about naked on the leaves
but normally protect themselves by making a case
out of bitten-off pieces of
leaf. Sometimes one piece
of leaf is attached by silk
to the underside of a whole
leaf ; the larva lives inside
this shelter protruding its
head to eat. More usually
the case is made of two
pieces of leaf fixed together
with silk, and this structure is portable. The
young caterpillars breathe
through their skin and
are easily wetted. After
several moults, however,
the skin becomes covered
with hairs which make the
surface unwettable. From

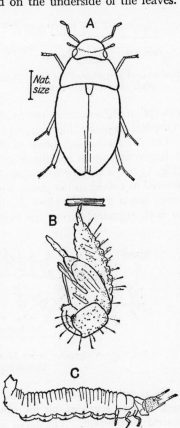

FIG. 131.—*Philydrus*.

A. Adult. After Joy.   B. Pupa. After
Brocher.   C. Larva.   After Brocher.

then on the caterpillar is
surrounded by air inside its case. Before pupating the caterpillar fixes its case to the water-plant with silk and closes both
ends. The pupa is also surrounded by air. According to some

the animal pupates inside its case above the level of the water, others say that it pupates under water. The adults emerge and are to be found on the wing in June and July. They

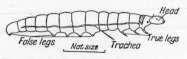

FIG. 132.—Caterpillar of the Moth (*Nymphula (Hydrocampa) nympheata*).

are common near water, over the surface of which they fly at dusk and also quite often in the daytime. The whole life-cycle probably takes a year; during the winter the young caterpillars hibernate in the mud at the bottom of the pond. *Nymphula nympheata* (Fig. 132) is perhaps the commonest species.

On the leaves and stems of the Water Soldier (*Stratiotes*) a larger and more handsome caterpillar is sometimes found (Fig. 133). This used to be called *Paraponyx*, but it is now considered to belong to the genus *Nymphula*. It grows to 2·3 cm. long, has a greenish-yellow transparent body, and attached to each segment, except the first thoracic, are a number of

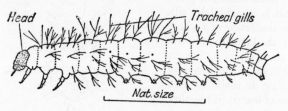

FIG. 133.—Caterpillar of the Moth (*Nymphula (Paraponyx)*).
After Brocher.

tracheal gills. Usually there are several tufts of gills on the back of each segment, on the sides, and several more almost ventral in position. The larva is not surrounded by air in its shelter but by water; periodically it undulates its body which changes the water inside its retreat so that the gills can obtain more oxygen. The caterpillar has spiracles but these are closed. The pupa is formed under water; it is surrounded by air inside a cocoon, and now the spiracles open, so it is able to breathe air directly.

Another small caterpillar is commonly found among the floating masses of duckweed. This is *Cataclysta lemna ;* it makes its shelters out of several leaves rather more in the manner of a caddis larva. It is very like species of *Nymphula* in appearance and habits.

One other caterpillar is found on a variety of water-plants such as the Arrow-head, *Ceratophyllum*, etc. This belongs to a different family from the others and is called *Acentropus niveus*. The adult is a white moth which appears in July. The females enter the water and lay their eggs in rows on water-plants. The larvæ are rather like those of *Nymphula*, only smaller ; when full grown they measure 1·2 cm. They are yellowish-white in colour with a darker head and thorax, the sides are brownish. They live surrounded by air and breathe through their spiracles. Like the other aquatic caterpillars they build a house out of leaves and silk. Pupation takes place several metres under water. A curious feature about this species is that there are two kinds of female moths, one has wings and is able to fly about, while the other has only rudimentary wings and remains under water attached to plants.

Order Hymenoptera

The best-known non-aquatic members of this order are the Bees, Ants, and Wasps. As well as these very familiar insects the order contains huge numbers of other creatures, many of which are parasites like the ichneumon-flies and the chalcids. The only fresh-water species are small Hymenoptera whose larvæ parasitise either the eggs or young stages of other aquatic insects. There are probably quite a number of species, but their habits are little known.

The characteristics of the Hymenoptera may be summed up as follows : the adults have two pairs of membranous wings of which the second pair are smaller than the first. On each side the smaller member of each pair (i.e. the hind wing) is locked to the hind border of the front wing by means of hooks, so that the two pairs beat together. The veins on the wings are much reduced in number and are sometimes absent. The mouth-parts are adapted for biting (with the mandibles) and often for sucking as well (in which case the labium is modified and often elongated). The abdomen is usually constricted just behind the thorax to form a " waist," the first abdominal segment being actually fused with the thorax. The females possess an egg-laying apparatus or ovipositor at the hind end of the body which can be used for the various purposes of piercing plant tissues, sawing, or stinging. The larvæ are usually legless, with a well-developed head. Pupation takes place in a cocoon.

Super-family Ichneumonidæ.   Ichneumon Flies

Genus Agriotypus.—Members of this genus lay their eggs inside the cases of caddis larvæ. The adult ichneumon-fly (Fig. 134A) swarms over the surface of streams in spring. The females crawl down plant stems into the water and then creep under stones in search of caddis larvæ of the family Phryganidæ (p. 186) ; she then lays her eggs inside the case. On hatching, the young larvæ of the ichneumon (Fig. 134B) devour the tissues of the caddis larva, but do not eat any of the important structures, such as the nervous system of their host, until the caddis has prepared its case for pupation, by fixing it down to some solid object with silk threads, and closing both ends. The remainder of the caddis is then eaten and the parasites pupate inside the caddis case, each in their own silken cocoon. There they remain until the following spring when they emerge. Parasitised cases of the caddis are easily recognised because

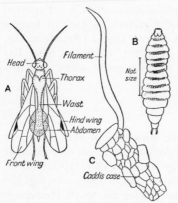

FIG. 134.—An aquatic Ichneumon Fly (*Agriotypus*).

A. Adult.   B. Larva.
C. Parasitised caddis case.
After Klápálek and Brocher.

they have a long thread attached to one end which is made by the parasite (Fig. 134c).

Some other nearly related species of the family *Braconidæ* parasitise aquatic fly larvæ, eating the tissues of their pupæ.

Super-family Chalcidoidea

All the species in this super-family are small insects and two genera are known to be aquatic.

Genus Polynema.—Species of this genus are about 1 mm. long (Fig. 135). The females lay their eggs inside those of other insects such as Agrionid dragon-flies (p. 126) or *Notonecta* (p. 141). One egg only is usually laid inside that of the host ; from this a tiny larva hatches which proceeds to eat

 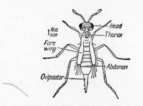

FIG. 135.—*Polynema natans*, side view. After Lubbock.

FIG. 136.—*Prestwichia aquatica*, adult female. After Brocher.

up all the yolk and developing embryo inside the host's egg. When it has finished this material it pupates in the egg-shell, and emerges as an adult in about ten to twelve days. The adult *Polynemas* swim under water using their wings (Fig. 135).

Genus Prestwichia.—The species *P. aquatica* parasitises the eggs of insects, including the bugs *Notonecta* and *Ranatra* (p. 145) and the beetles *Dytiscus* and *Pelobius* (p. 161). *Prestwichia* is also only 1 mm. long ; it lives under water as an adult and is the most really aquatic of all the Hymenoptera (Fig. 136). The adults are found in summer ; the males have no wings, while the females have only small non-functional ones ; they swim in water using their legs. Several adults emerge from each egg of the host species.

Super-family Proctotripoidea

Genus Diapria.—Species of this genus attack the larvæ of flies, including that of the aquatic rat-tailed larva (*Eristalis*).

The Caddis-flies. Order Trichoptera

The adult caddis-fly, which is a moth-like insect found flying about near water, is very much less familiar than its larval stage, the caddis worm, which is aquatic. The adults have very long many-jointed antennæ, a vestigial first pair of jaws, two pairs of elongated membranous wings which are covered with hairs and in which the venation is not distinct on first sight ; these are held at rest so that they form a roof to the abdomen (Fig. 137). They are not very strong fliers and the

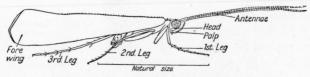

FIG. 137.—An adult Caddis-Fly (*Leptocerus*).

wings and body are always rather dull coloured, greys and fawns being the prevailing shades.

The eggs are laid by the females in spring and summer, most frequently under water, attached to stones or plants. They are either laid surrounded by a material which swells up into a jelly mass in contact with water (Fig. 138c) or they are laid in flat masses cemented together. The larvæ on hatching often make some kind of portable protective case or house in which to live. This may be made of plant material, stones, sand grains, or silk only ; or else they construct fixed silken tunnels. When foreign material is used for the case this is always cemented together and lined throughout with silk, which is produced by the salivary (labial) glands as it is in caterpillars of butterflies and moths. The case is usually a good deal larger than the larva, and it is added to as the larva grows bigger. The shape and material of the case help greatly in identifying the various genera. Sometimes the larva does not make a case at all, until it has moulted

several skins ; in a few rare instances the larva lives free
all its life and never builds a case, silken net, or tunnel.

Caddis larvæ are easily recognised even if they have no
case, or if they have accidentally lost it. In all of them the
head is well chitinised, the antennæ are generally very small
and the mouth-parts are of the biting type. Often the head
is patterned with a geometrical design in green and brown,
or yellow and brown. The thorax has the first segment only,
or sometimes the first two, chitinised in those forms which
build cases of foreign material. The legs are of unequal
length, the first pair are the shortest and stoutest, while
either the second or the third pair may be the longest. The
abdomen is nearly always very clearly divided into nine
segments. In all large species the abdomen bears gills of
a tubular type with tracheæ inside, and these are arranged
in various ways. The last abdominal segment always carries
a pair of jointed appendages ending in hooks, which are very
prominent curved structures in some species. These are
used in case-building forms to grip hold of the case while the
animal walks or swims forwards with its legs. Those living
in silken tunnels use them to grip the silk threads, or to hold
on to rock surfaces when they leave their retreats. This pair
of abdominal appendages, anal appendages as they are sometimes called, are quite characteristic of caddis larvæ. The
only other aquatic creature to possess anything like them is
the larva of the whirligig beetle *Gyrinus* (p. 166).

Caddis larvæ can be divided into two types, the rather
sluggish, portable case-building kind, which are generally
vegetarians, and the much more active variety living free or
in silken tunnels which are generally at least partly carnivorous. The sluggish vegetarians have much softer more
grub-like bodies, the abdomen being fatter. The first
abdominal segment bears three fleshy protuberances which
push against the side of the case when the animal has its head
and thoracic legs extended forwards. The anal appendages
are relatively small (Fig. 140A). The other more active type
is characterised by having a slimmer less fleshy body with
the abdomen tapering towards the last segment. There
are no protuberances on the first abdominal segment, while

the appendages on the last segment are often very well developed and bear large recurved hooks.

Caddis larvæ always pupate inside a shelter of some kind. Case-building types make use of the larval case to which they often add extra material to block up the entrances at either end before attaching it with silk to some support. Those forms which live free or in silk nets leave their old haunts before pupation to construct a fixed shelter on the underside of stones ; this is often made of small stones or sand grains and is lined with silk. Inside the shelter the larva normally constructs a silken cocoon, the threads of which are cemented together to form a substance rather like flexible chitin in appearance. Inside the cocoon the larva sheds its last skin and becomes a pupa. In this stage some of the adult features, such as the long antennæ and wings, become apparent for the first time, though of course these structures have been developing internally for some months. On the abdomen the larval gills persist so that the pupa is able to breathe through them. After a few weeks, or in some cases the whole winter, as a pupa, the animal is ready to emerge as an adult. The pupa possesses huge strong mandibles with which it bites its way out of its shelter ; it then swims to the surface of the water using its middle pair of legs which are very long and have swimming hairs (Fig. 144B). If it is living in running water the pupal skin is shed almost immediately the pupa reaches the surface and the adult flies off into the air. In still water species on the other hand the caddis is more leisurely about its emergence. The life-history appears to last about a year. The larva may pass the winter in a semi-dormant condition at the bottom of a pond pupating in early summer, or the winter may be passed in the pupal stage, the adult emerging in spring.

The following is a description of the larvæ of common genera of caddis-flies found among the families occurring in Britain :—

Family Phryganidæ

The larvæ (Fig. 138A) of this family are fairly large when full grown, about 20-40 mm. long. They build cases (Fig. 138B) of plant material, such as pieces of roots, leaves, or reeds. Each piece of plant is cut into a rectangle, and then a case is made out of these by arranging them in a spiral to form a cylinder, open at both ends, the spiral twisting to the left. In young larvæ the case may be slightly conical. The larvæ may easily be induced to leave their cases by poking them from the hind end; if the case is then removed and the larva supplied with some leaves it will quickly make a new case. Normally the larva comes out of the front end of the case when poked from behind, clings to the side and crawls to the far end, where it enters the case again. It can be made to repeat this action several times, after which it will often go on crawling out and in on its own! When the larva is removed from its house it will be found to have the following features: a well-chitinised head and first thoracic segment,

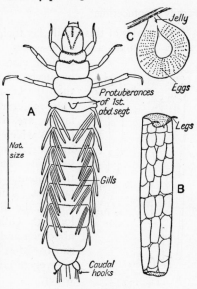

Fig. 138.—*Phryganea.*

A. Larva.
B. Larval case made of pieces of leaves.
C. Egg mass.

a broad fat abdomen ending squarely, small anal appendages with hooks, three large protuberances on the first abdominal segment and a number of simple filiform gills attached to the sides of the abdomen (Fig. 138A). These larvæ are mixed feeders eating other insects and molluscs as well as plants. A few species are almost entirely vegetarians.

When ready to pupate the larva fixes its case to some support with silk and then partly covers the open ends. The animal pupates inside. The adult usually emerges in a few weeks.

The female lays her eggs in summer in gelatinous ropes which are attached to water-plants (Fig. 138c). The larvæ are found in still water such as ponds and ditches where there is much plant life. Several species of the genus *Phryganea* are common in Britain of which *P. grandis* is shown in Fig. 138.

Family Leptoceridæ

The females of this family lay their eggs in round gelatinous masses on stones and water-weeds. The larvæ build narrow conical cases which may be straight or curved in outline. They are bigger than the larva and are made of vegetable material or sand grains. The larvæ are slender with cylindrical bodies which have their broadest part in the region of the second thoracic segment. The head is elliptical and the antennæ are longer than in other families. The legs are of very unequal length, the third pair being much the longest. Gills may be present on the abdomen and the anal appendages are short. Most of the larvæ in this family are vegetarians and are found in still water.

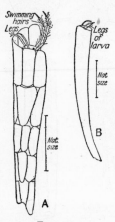

FIG. 139.

A. *Trianodes*, a swimming Caddis with case made of pieces of leaf.
B. A *Leptocerid* case of sand grains.

Genus Trianodes.—Case of larva (Fig. 139A) made of pieces of vegetable material arranged in a spiral as in the family Phryganidæ 20-30 mm. long. The larva is much smaller than its case, and it swims actively with its third pair of legs which are specially long and covered with hairs. It is a vegetarian. Before pupation the case is reduced in size, and fixed to plants. The eggs are found on the undersides of floating leaves ; they are laid in a gelatinous disc.

Genus Leptocerus.—The larval case (Fig. 139B) of this genus is conical and fairly strongly curved. It is about 11-17 mm. long, usually made of sand grains.

Genus Setodes.—The larvæ are small, only 5-8 mm. long. They live in conical cylindrical cases which are wholly made of a sticky secretion in species living in still water. Some live in running water and then the cases are made of sand. The larvæ have the first two thoracic segments chitinised, long third pairs of legs which are used for swimming in pond-dwelling species, three projections on the first abdominal segment well developed, while the gills and the anal append-ages are very short.

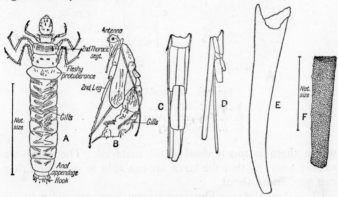

FIG. 140.—Limnophilid Caddis Larvæ.

A. Larva of *Limnophilus politus*. After Rousseau.
B. Pupa of *L. flavicornis*. After Rousseau.
C. Case of *L. xanthodes*, made of pieces of leaf.
D. Case of *L. decipiens* made of plant material.
E. Case of *L. vittatus* made of sand and plant particles.
F. Case of *Stenophylax* made of sand grains.

Family Limnophilidæ

This is a large family in which the larvæ make their portable cases out of sand grains, stones, vegetable material, or mollusc shells. The case is cylindrical, less frequently conical. The larvæ are fat with soft bodies rather like those of *Phryganea*, but the second thoracic segment bears a chitinous plate on its back which is clearly divided into two by a median division ; the second pair of legs is the longest.

Genus Limnophilus.—There is a large number of species many of whose larvæ make cases of a characteristic shape (Figs. 140,

141), though this character alone is not always sufficient to identify a species. They live in stagnant or slow-flowing water

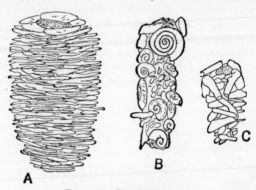

Fig. 141.—*Limnophilus* cases.

A. *Limnophilus stigma.*
B. *L. flavicornis.* After Rousseau.
C. *L. rhombicus.*

where there is much dead plant material. The cases are usually heavy so that the larvæ are not able to swim and they can only crawl about.

Genus Stenophylax.—The larval case is a straight cylinder made of sand grains (Fig. 140F).

Genus Anabolia.—The larval case is 16-30 mm. long and is made in the young larva of plant stalks placed lengthwise to form a cylindrical structure ; in older larvæ grains of sand and snail shells are used as well as plant material.

Family Molannidæ

There is only one British genus, *Molanna*, whose larva is found in the still water of lakes and large ponds. The case is quite characteristic (Fig. 142), made of sand grains in the form

Fig. 142.—Case of *Molanna*.

of a flattened tube with a kind of hood arrangement at the head end.

Family Hydroptilidæ

In this group the smallest caddises are to be found. The eggs are laid in plaques on plants. The larvæ make cases which are not tubular, but flattened. They are usually made of silk only and are open at one end (Fig. 143). The larvæ are only a few millimetres long ; they are very narrow at the head end but the abdomen is broad, the greatest width of the body

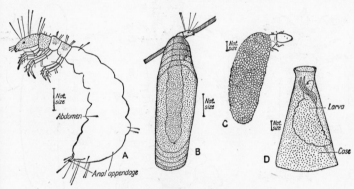

FIG. 143.—Hydroptilid Caddis Larvæ.

A. *Agraylea* larva.
B. *Agraylea* larva in case.
C. *Hydroptila*. After Klápálek.
D. *Oxyethira* larva in case.

being between the fourth and sixth abdominal segments. There are no gills and the anal appendages are very small. The young larvæ do not live in cases until after several months ; the diet is vegetarian. Before pupation the case is attached to some support and the larva forms a pupa inside, using the case as a cocoon.

Genus Agraylea.—Case about 4·5-6 mm. long (Fig. 143A, B), fairly common in still water.

Genus Hydroptila.—Case 3-4 mm. long (Fig. 143C), made of sand grains as well as silk, common.

Genus Oxyethira.—Case shaped like a flask (Fig. 143D), about 5 mm. long. Not very common.

Family Rhyacophilidæ

In this family the eggs are laid in round cemented masses on the undersides of stones in streams. The larvæ are carnivorous creatures found in the rapidly flowing part of rivers, streams, and mountain rivulets. The larvæ do not always form cases, but if they do they are of crude construction. The body of the larva is narrow in proportion to its breadth and they are active animals with well-developed anal appendages.

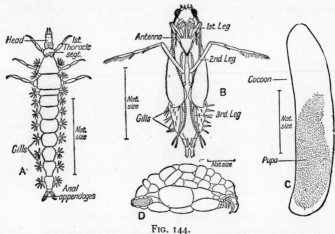

FIG. 144.

A. *Rhyacophila* larva.   B. *Rhyacophila* pupa.
C. *Rhyacophila* cocoon.   D. *Agapetus* larva in case.

Genus Rhyacophila.—Members of this genus are found as larvæ on the under surfaces of stones in rapidly flowing streams. They live free and do not build a case or shelter of any kind. The larvæ are 9-24 mm. long with a green and brown body. They are very active and can creep about quite quickly, using their long anal appendages for gripping stony surfaces. They have a small head, the first thoracic segment in the shape of a quadrangle, and the abdominal segments are elliptical in outline (Fig. 144A). The gills when present are in bunches on either side of the abdomen, but in some species these are lacking. The anal appendages are large with curved hooks.

The larvæ live for quite a long time in an ordinary aquarium in spite of the fact that they are used to well-oxygenated water. The larva is full grown in the late autumn ; it constructs a pupal shelter by cementing small stones together on the underside of a large stone ; inside this it makes a brown cocoon of a leathery consistency which is transparent and the shape of a curved cylinder closed at both ends (Fig. 144c). Pupation takes place inside the cocoon. The pupa (Fig. 144b) bites its way out of the cocoon and shelter before the adult emerges, and swims to the surface where it rapidly sheds the pupal skin and flies off. Species of *Rhyacophila* are very common in upland streams, and their bright green and brown bodies make them easy to identify.

Genus Agapetus.—The larvæ of this genus live in stone cases in mountain streams and rivulets. The case is a curious shape, being flat on one side and curved on the other, with two openings on the flat side (Fig. 144d). The length of the larva is about 4·5-7 mm. Before pupation the stone case is attached to a larger stone and the larva pupates inside. Species of this genus are very common in some places, numerous individuals being found under one stone.

Family Philopotamidæ

Genus Philopotamus.—The eggs are laid on stones in mountain streams. The larva is an active creature with a narrow body 22 mm. long. It has no gills on the abdomen ; the anal appendages are large, and the insect does not make a portable case but lives in a retreat which it spins of silk threads. It leaves this before pupation and makes a protective covering of stones inside which to pupate.

Family Hydropsychidæ

Genus Hydropsyche.—This is another genus which is nearly always found in swift streams during its larval and pupal period. The eggs are laid on stones under water in round cemented masses. The larvæ are very active (Fig. 145a) ; they are 10-20 mm. long with a body of nearly uniform width gradually tapering towards the hind end. There are gills on the abdomen in all except very young larvæ, and there are four anal gills as well ; the anal appendages are very large. The larvæ do not build cases but live in silk nets on the undersides of stones. These nets are made with a wide opening which leads into a silk tunnel ending blindly. The larva remains in the tunnel which faces upstream. Any animal or plant material which is caught in the net is seized by the larva and eaten ; the diet is therefore mixed and the larva really is a filter feeder using its net as a sieve. The net is made of strong silk threads, and the long bristles at the end of the abdomen are used as cleaning brushes to keep the net clean. Before pupating the larva leaves its silk net to make a stone shelter (Fig. 145B). Species of this genus are common in streams and sometimes in ponds.

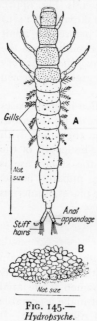

Gills

A

Nat. size

Anal appendage

Stiff hairs

B

Nat. size

FIG. 145.—
Hydropsyche.

A. Larva.
B. Pupal shelter

Family Psychomyidæ

In this family the eggs are also laid in cemented masses attached to stones or other objects in still or running water. The larvæ do not make cases, but they make galleries of sand or stones on the undersides of submerged objects. They have very long anal appendages and no lateral gills.

Genus Tinodes.—The larvæ of this genus are long and thin, measuring about 8-11 mm. by 1·2 mm. They have an elliptical flattened head and long hairs on the front border of the rectangular first thoracic segment. The legs are short and strong. All the abdominal segments are the same size except the last, and the anal appendages are fairly large with two joints. There are five filiform anal gills. The larvæ make galleries out of silk, sand, and plant débris which is all stuck together by a brownish-grey secretion. These galleries are about 30-40 mm. long and form a tortuous tunnel. The larvæ are apparently vegetable feeders. They are found in slow-flowing streams and also at the edges of lakes. The pupal shelter is made of sand and encloses a cocoon.

Genus Psychomyia.—This is very similar to the previous genus in habits, but the larvæ are much smaller being only about 4·5 by 1 mm. Their body is fattest near the first to third abdominal segments ; the third pair of legs are the longest and there are no hairs on the body. The pupal shelter is made of sand grains. Species of this genus are found in both still and running water.

Family Polycentropidæ

This is another family whose members are carnivorous web-spinners. The larvæ are fairly large with a slightly flattened body. The head has very small antennæ, the legs are short, the middle pair being slightly the longest. There are no gills on the side of the abdomen but there are five anal gills on the last segment. The anal appendages are large with two or three joints and long terminal hooks. The larvæ do not make cases but live in silk webs. A pupal retreat is built of stones, sand grains or pieces of plants ; this is lined with silk and encloses a grey transparent cocoon which is open at both ends.

Genus Plectrocnemia.—The larva is large, about 22 mm. long by 3 broad. It has a dark yellow-brown head and the body is broadest about the second abdominal segment. The silk tunnels have a wide opening narrowing to a cul-de-sac. In this net animals and plants which are carried down stream are caught, the larva then seizes and eats them. Should the stream dry up the larvæ creep into crevices between stones until the stream is once more full of water. The pupal retreat is flattened against the underside of a stone ; it is made of sand grains, small stones or vegetable débris. This genus is common in swift-flowing streams.

Genus Polycentropus.—Species of this genus are smaller than the last, the average length being about 12 mm. The body is uniformly broad with short hairy legs and large two-jointed anal appendages with claws almost at right angles to the last

joint (Fig. 146A). The larvæ are found in rivers and near the
edge of lakes, sometimes many live together in societies. The
silk net is something like the shape of a swallow's nest with
the opening facing the current (Fig. 146B, C).

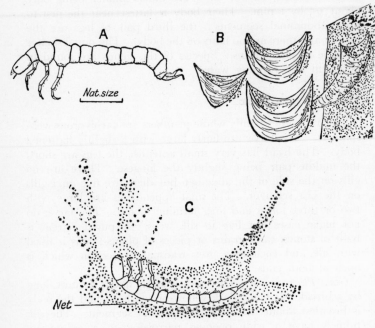

FIG. 146.—*Polycentropus.*

A. Larva.   After Brocher.
B. Silk nets made by larva.   After Wesenberg-Lund.
C. Net made by larva in glass aquarium.

Genus Holocentropus.—The larvæ are about the same size
as those of *Polycentropus* (i.e. 10-13 mm. long), but they are
broadest in the middle of the abdomen. The long three-
jointed anal appendages are hairy and have the hooks almost
at right angles to the last joint. Species of this genus are
found only in still water. They live under the large floating
leaves of water-lilies, inside silk nets which they attach to the

floating leaves. The pupal shelter is made of various materials, but usually the larva spins pieces of plants together.

To identify genera of larvæ see

1. Les Larves et nymphes aquatiques des insectes d'Europe, by E. Rousseau. Vol. I, Bruxelles, 1921.
2. Die Süsswasser Fauna Deutschlands, edited by A. Brauer. Tricoptera.
3. Larvæ of the British Tricoptera, by N. E. Hickin (1946). Transactions of the Royal Entomological Society of London, **97,** 187.

Order Diptera. The True Flies

This order contains a very large number of insects, of which a fairly large proportion live in water for part of their lives. The adults (which are never themselves aquatic) may be easily recognised because they have only a single pair of wings which are transparent, membranous, and usually clearly veined with grey or black. The second pair of wings are reduced to small sensory knobs which are used as balancing organs. These balancers can be seen easily with a hand lens in a large fly like a bluebottle. The mouth-parts of adult flies are usually enclosed in a proboscis (mandibles are lacking except in the females of some blood-sucking flies ; the labium forms the proboscis). They are designed for sucking up liquid food, and for this purpose some part of them forms a tube. In some forms, such as the gnats and midges, the jaws are adapted to pierce as well as suck. The compound eyes on the head are often very large ; in the males of certain species the two eyes are so big that they are continuous across the head. The antennæ are either long, many jointed and furry, or very short consisting of a few broad joints. The three thoracic segments are fused together so that in a dorsal view they appear as one. The legs are of the ordinary slender running or walking type, usually with five joints to the tarsus, the last one having claws. In the abdomen some segments at the front end are fused, while others at the hind end are telescoped internally. In the crane-flies it is possible to see segments three to eleven, but in most other forms it is only possible to make out four or five segments without dissection.

All flies have a complicated life-history, involving the

stages of egg, larva, pupa, and adult, the first three of which stages may be passed through in water. The females of those species which have a partly aquatic life, lay their eggs in masses on the water surface, or in jelly attached to plants, or they scatter them at random over the water surface. The larvæ of *Diptera* have various shapes but they are normally grub-like ; they *never* have proper jointed legs on the thorax, though they may have fleshy false legs on the thorax and abdomen. The last abdominal segment may be provided with " gills," a breathing siphon, or various projections. Some families have larvæ with a well-developed head with antennæ and mouth-parts, while others have no real head visible at all, because it is retracted inside the thorax ; the jaws are then reduced to a pair of mouth-hooks. The skin is nearly always soft on the thorax and abdomen, usually it is transparent, and either white or greyish in colour. For this reason the tracheæ generally show up very clearly as silver or black tubes running from the spiracles to all parts of the body. In mosquito larvæ they are particularly easy to see, and some of the very fine branches ending in individual muscles and other organs may be made out under the low power of the microscope. Many larvæ have a pair of large spiracles near the end of the body through which air is breathed by the tail end, or a prolongation of it, being placed above the water surface. Others breathe the dissolved air through their skins.

The larvæ may live freely in the water and be active swimmers, a very few float in the surface water of lakes, while many more live in mud or among thick plant growth. A large number of midge larvæ make sand and mud tubes in which they live, other forms are amphibious, creeping about under water among stones and mud, or coming out on to damp ground. Really aquatic larvæ usually pupate in the water. A number of these pupæ are active swimmers. In some families the pupa is formed inside the last larval skin which is specially hardened as a loose protective covering ; this is called a *puparium*. The pupa in such cases is immobile and none of the adult structures such as wings and antennæ are visible through the puparium (Figs. 156c, 157b).

The classification of the *Diptera* is a difficult subject, but they may be divided into two large groups by slight differences in their life-histories. The first group is called the *Orthorrhapha*, and in them the adult emerges from its pupa through a dorsal longitudinal slit in the skin of the latter. In the *Cyclorrhapha* the pupa is enclosed in a puparium, and the adult emerges by one end of the puparium splitting off as a cap. Each group has two sub-divisions in which there is much difference in the characters of the larvæ, so a brief description is included here.

Orthorrhapha

(*a*) *Nematocera*.—Larvæ have a well-developed head with antennæ and biting mandibles. The pupa is free and in aquatic forms often active. The adults have long many-jointed antennæ which usually exceed in length the head and thorax. The principal families with some aquatic species are the crane-flies or daddy-long-legs (*Tipulidæ*); the moth-flies and sand-flies (*Psychodidæ*), family Dixidæ ; the gnats and mosquitoes (*Culicidæ*), the midges (*Chironimidæ*), and the black-flies (*Simulidæ*).

(*b*) *Brachycera*.—The larvæ have an incomplete head which is generally retractible into the first thoracic segment. The pupa has its legs free from the body. The adults have antennæ which are shorter than the thorax. The principal families with aquatic species are the *Stratiomyiidæ* and the horse-flies (*Tabanidæ*).

Cyclorrhapha

(*a*) *Athericera*.—The larvæ have a vestigial head and the pupa is enclosed by the last larval skin, so that it is immobile. The adults have only three joints to their antennæ. The chief families with aquatic species are the Hover Flies (*Syrphidæ*) and the *Anthomyidæ*.

(*b*) *Pupipara*.—No aquatic species.

A. *Nematocera*.—The following are the common forms with aquatic stages in their life-history :—

Family Tipulidæ, The Daddy-long-legs or Crane-Flies

These flies are among the largest found in the order, and their larvæ are fat, worm-like creatures, rather dirty white in

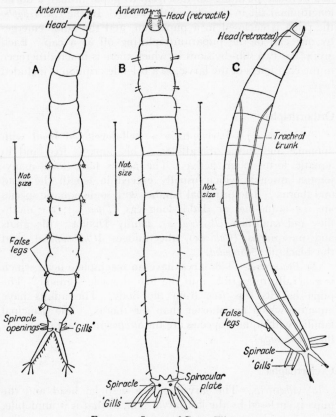

FIG. 147.—Larvæ of Crane-Flies.

A. *Dicranota.*   B. *Tipula.*   C. *Pedicia.*

colour, which may be several centimetres long. They are found in damp situations near streams, or among the mud and stones in a shallow stream bed or among the mud in ponds. The head of the larva is well developed, but in some genera it can be retracted at least partly into the first thoracic segment. The abdomen has fleshy false legs in some genera and the last segment has a pair of large spiracles as well as three pairs of gill-like outgrowths. The appearance of this last segment is a useful guide to the identification of the genus.

Genus Dicranota.—The larvæ live among mud in streams and ponds ; they are carnivorous and eat *Tubifex* worms (see p. 53). They are easily recognised by their five pairs of cylindrical false legs on segments three to seven of the abdomen ; these each end in a circlet of hooks. The full-grown larva measures 2 cm., and it probably takes two or three months to reach this stage (Fig. 147A). The pupa is much smaller than the larva, about 1 cm. long with a pair of short breathing horns on the first thoracic segment.

Genus Tipula.—The larvæ are large, reaching a length of 3 cm. or more when extended. The head is retractile ; the last abdominal segment has a spiracular plate which is quite characteristic, in that it is almost rectangular with either six or eight surrounding lobes (Fig. 147B). The pupa has long breathing horns. Common in damp mud and under stones in shallow streams.

Genus Pedicia.—The larvæ are whitish, 4 cm. long when full grown. The head is very retractile. The abdomen has four pairs of false legs on segments four to seven (Fig. 147C).

Family Psychodidæ. Moth-Flies

The adults of this family are very small flies which have bodies and wings covered with conspicuous hairs. The larvæ of some genera are aquatic and are often found in sewage beds. The eggs are laid in jelly masses from which the larvæ hatch in about a fortnight. They have long cylindrical bodies with no false legs ; the head is not retractile and the thorax has a pair of spiracles in addition to there being a spiracular opening at the tip of a respiratory tube which is at the posterior end of the body. Most forms have hardened plates of skin on the dorsal sides of some of the segments. The pupa has its legs free but they are hidden by the wing covers.

Genus Pericoma.—The larva has a plate of hard skin on each segment covered with sensory hairs.

Genus Psychoma.—The larva has a plate of hard skin only on the posterior segments.

Family Dixidæ. The Dixa Midges

This is a very small family with only one genus, that of *Dixa*. The adults are rather like gnats, but the long antennæ

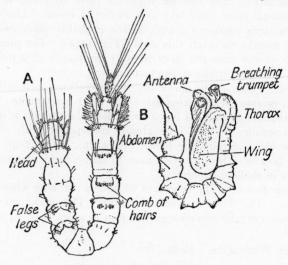

FIG. 148.—The *Dixa* Midge.

A Larva, ventral view.
B. Pupa. Both after Johannsen.

are hairless. The aquatic larvæ are found in ponds which contain a good deal of vegetation, and they live near the surface among floating leaves. The body is a little more than half a centimetre long in a full-grown specimen and it is characteristically held bent double (Fig. 148A) so that the larvæ are easily recognised. They possess a well-developed head with mouth brushes like mosquito larvæ (see p. 205). The first thoracic segment has very long hairs on the anterior border. There are two pairs of false legs ending in a comb of hairs on the first and second abdominal segments, while the fifth, sixth, and seventh segments have each a pair of combs. On the dorsal surface at the tail end there is a pair of spiracular openings which are surrounded by hairs to prevent their being wetted. There is also a conspicuous pair of side projections from the end of the eighth abdominal segment which are covered with hairs and the extreme end of the tail has several very long hairs. The larva feeds on microscopic particles in the water which it collects by whisking its mouth brushes to and fro. The larva normally has its tail end out of the water, but it can swim if completely submerged by jerking its body from side to side. The pupa (Fig. 148B) rests motionless at the surface with its abdomen curled up so that the tail is on a level with its eyes.

Family Culicidæ. The Gnats and Mosquitoes

The adults of this family are well known from their unwelcome attentions to human beings ! They are very slender flies with a proboscis adapted to pierce as well as suck. In blood-sucking species it is only the female which has this habit, and she is often unable to lay any eggs without having had a meal of blood. The antennæ are long and in the males very much covered with long hairs ; the females have far fewer hairs, and this feature forms an easy method of distinguishing the sexes. The larvæ and pupæ are all aquatic, both being very active swimmers. It is on account of the larvæ and pupæ living in water that the adults are never found very far from slow-flowing or stagnant water, though for some species a small water hole in a tree-trunk or a rain barrel is sufficient. The family is divided into the Phantom midges (*Chaoborinæ* or *Corethrinæ*) and what are usually called mosquitoes (*Culicinæ*, e.g. *Culex*, *Anopheles*, etc.)

The Phantom Midges (*Chaoborus*).—The larva of these is characterised by having prehensile antennæ with which it seizes its food. It is carnivorous, living on other insects and crustaceans. The long cylindrical body is completely transparent, the only easily visible structures being the black eyes and two pairs of air-filled hydrostatic organs one at either end of the animal (Fig. 149A). The larva normally lies motion-less suspended horizontally in the water. When any suitable food swims within reach, it is seized by the antennæ. The last segment of the abdomen bears a stiff comb of hairs on the ventral side called the rudder ; behind this is a pair of strong hooks, and at the very tip of the abdomen is a group of retractile gill-like appendages. The pupa has well-developed breathing trumpets and a pair of flexible paddles on the last abdominal segment (Fig. 149B). The adults are not blood-suckers ; the females lay their eggs on the surface of water in flat gelatinous discs which may contain 100 eggs arranged in a spiral. The larvæ are found in small clear ponds as well as in large lakes where they occur near the surface, deep down or even in the mud at the bottom ; they are rather local in occurrence. *

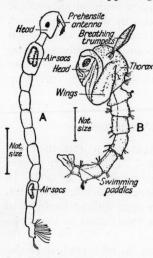

Fig. 149.—The Phantom Midge (*Chaoborus* late *Corethra*).

A. Larva.
B. Pupa. After Johannsen.

The Mosquitoes or Biting Gnats. Sub-family Culicinæ.—There are some thirty British species of biting gnat. The sub-family Culicinæ is divided into two sections or so-called "Tribes," the Anophelini (genus *Anopheles*) and Culicini (genera *Culex, Aëdes, Theobaldia, Orthopodomyia, Tæniorhynchus*).

Genus Anopheles. Five British species. These include a vector of malaria, a disease which was formerly common in the fens and south-east of England when *A. maculipennis* was more numerous. It should be noted that the mosquito only transmits the disease, becoming infected when it (the female of course) bites a malaria patient ; she then infects other people when feeding again, after the malaria parasite has developed in her body. So *Anopheles* alone will not cause malaria, they only make its transmission possible. These mosquitoes are

* To identify genera or species of Chaoborinæ, see Zur Kenntnis der Larven und Puppen der Chaoborinæ (Corethrinæ von Fritz Peus (1934). Archiven für Hydrobiologie, **27**, 541.

easily distinguished in the adult stage, where the female rests
with her head, thorax, and abdomen in a straight line at an
angle to the surface, so that she is said to " stand on her head "
(Culicini rest with their bodies more nearly parallel to the
surface).

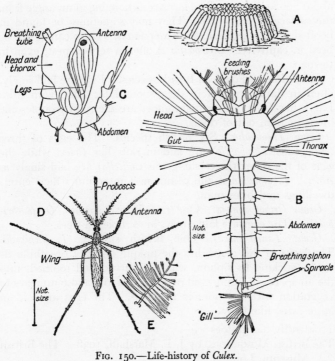

FIG. 150.—Life-history of *Culex*.

A. Egg raft.   B. Larva.   C. Pupa.   D. Adult female.
E. Antenna of male.

Anopheles eggs are laid singly on the surface of the water ;
they have " floats " at the side.   The larvæ are built on the
same plan as that illustrated (Fig. 150B) but the breathing
siphon is very short, and the larva normally lies parallel to
and just below the water surface.   Like other larvæ of this sub-
family, it feeds by whisking a pair of mouth brushes to and fro
in front of the head ; these filter tiny plants and particles of
organic matter out of the surrounding water and the material

thus collected is taken into the mouth. The pupa is like the *Culex* illustrated (Fig. 150c), with small breathing trumpets and a pair of fins at the end of the abdomen ; it is an active swimmer. *Anopheles* are rather tricky to breed in captivity.

Genus Culex. Three British species. Eggs are laid in rafts, about 200 at a time. The larva rests hanging at an angle from the surface of the water. They may sometimes be found in flooded basements and rain-water tanks. *Culex* can easily be bred in captivity. *C. molestus* is often a serious pest, even in towns.

Genus Aëdes. Fourteen British species. Eggs are laid singly, often in damp ground in depressions later filled with rain. Larva like *Culex*. *Aëdes* mosquitoes are the most troublesome biters of man out of doors.

Genus Theobaldia. Six British species. The eggs of three (sub-genus *Theobaldia*) are laid in rafts like *Culex*, while the eggs of the other three (sub-genus *Culicella*) are laid singly as *Aëdes*. *T. annulata* is the commonest mosquito which enters dwellings and bites man.

Genus Orthopodomyia. One rare British species, *O. pulchripalpis*. The eggs are laid singly, in rot holes in trees.

Genus Tæniorhynchus. One British species, *T. richiardii*. The larva of this genus has the unique habit of breathing its air by sticking its siphon, which is curved and toothed, into the air spaces in submerged plants. The pupa also attaches to vegetation with its thoracic trumpets. The eggs are laid in large rafts, like *Culex*.

To identify mosquitoes see

The British Mosquitoes, by J. F. Marshall, 1938. The British Museum, London.

Family Simulidæ. The Black-Flies

The black-flies belong to the genus *Simulium*. The adults are small greyish flies with very broad wings which have a specially distinct vein near the front border, the antennæ are very little longer than the head (Fig. 151A). The female attacks man, and gives a painful bite. Only the eggs of a few species are known, and these are laid in jelly masses or strings on stones and weeds near the edge of streams, where they will be wetted by the water. The larvæ on

hatching enter the water and attach themselves to stones,
or the leaves and stems of plants, in the fastest part of the
current. They are creamy-white or grey in colour with
a cylindrical body which swells out very considerably at the hind
end (Fig. 152). The head is provided with a very prominent

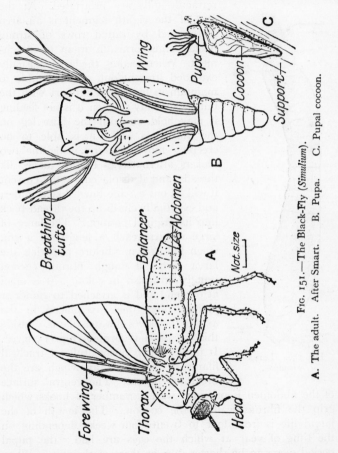

Fig. 151.—The Black-Fly (*Simulium*).

A. The adult.   B. Pupa.   C. Pupal cocoon.

pair of mouth brushes in addition to the ordinary jaws, and
with these small particles, mainly one-celled plants, are
whisked into the mouth. The thorax has one false leg (made
by the union of two) in the midline ; this ends in hooks.

and the arrangement forms a kind of sucker for gripping the surfaces of stones and plants. The segmentation of the abdomen is indistinct, the seventh is the largest segment and the eighth (the last) is very much smaller. The anus is on the dorsal side of the eighth segment; through its opening retractile gills are protruded. At the end of the eighth segment is an area surrounded by radial rows of strong hooks; this forms an organ of attachment. When feeding the larva is usually extended straight out, and it is able to maintain its hold on a support with the tail sucker only; when it is not feeding it may hold on with the false leg on the thorax as well; it is able to do creeping movements by using the two suckers alternately. It can make silk threads, and if displaced by the current it will let out a " life line " like a spider, and eventually climb up the thread back into its original position. The larvæ of large species reach a length of 2 cm. when full grown. Before pupating the larva forms a tough fibrous cocoon usually brownish in colour; it is open at one end and is attached to stones or plants; often they are placed in the axils of the leaves of water-plants. Inside this the pupa is formed (Fig. 151B, C). It possesses two curious tufts of tracheal filaments on the thorax which are the breathing organs. The ventral surface of the abdomen is provided with a number of hooks which grip the fibrous wall of the cocoon. The length of the larval life is from four to twenty-four weeks, depending on the time of year at which the eggs are laid; the pupal period seems to be shorter, two or three weeks only. When the adult is ready to emerge from the pupa it is surrounded by air which has been obtained from the water. As it emerges

Fig. 152.—Larva of *Simulium*.

the adult is carried up to the surface in a bubble of this air, and it is able to fly off at once.

The British Simuliidæ, with Keys to the Species in the Adult, Pupal, and Larval Stages, by J. Smart, 1944. Freshwater Biological Association of the British Empire. Scientific Publication No. 9.

Family Chironomidæ. The Midges

The adult flies are like gnats with the same kind of long antennæ, which are very hairy in the males and less so in the females. The jaws are not so well developed, however, and very few of them are able to pierce skin. The eggs are laid in jelly masses or ribbons in water except in a few genera which have land-dwelling larvæ. The aquatic larvæ nearly all have two pairs of false legs, one on the thorax and one on the last abdominal segment.

Genus Tanypus.—The eggs are laid in gelatinous blobs on leaves or stones under water. The larvæ (Fig. 153B) are common and are often found in large numbers together. They do not make any mud tubes of their own, but they are very often found inside those made by *Chironomous* midges on whom they feed. The body is slightly broader near the front end, tapering behind. The pair of thoracic false legs are fused together at their base. The false legs on the last abdominal segment are very long ending in stout hooks. There are four to six anal gills, and on the dorsal side of the ninth abdominal segment are a pair of thin backwardly directed papillæ ending in a group of long hairs. The pupæ are fairly active, resembling those of some mosquitoes ; they have long thin breathing trumpets on the thorax.

Genus Corynoneura.—The eggs of this genus are laid in cylindrical jelly masses about 2 mm. long. The larvæ are very small, fairly active, sometimes living in a loosely built tube made of débris. They may be recognised by their small size of only 2 or 3 mm., their long jointed antennæ, and very slender bodies. A common British species (Fig. 153A) is whitish banded with reddish-brown. The anterior false legs are fused at their base, and the posterior pair are very long. The second and third thoracic segments are fused together. The pupa may be recognised on account of its small size and lack of breathing trumpets.

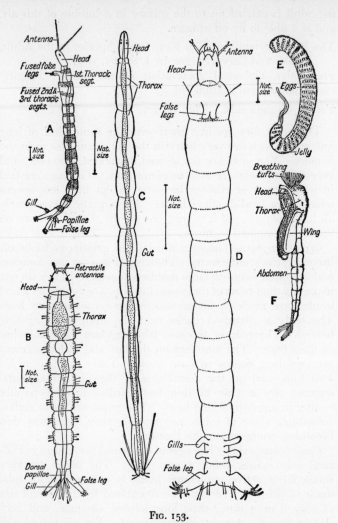

FIG. 153.

A. *Corynoneura* larva. B. *Tanypus* larva. C. *Forcipomyia* larva.
D. *Chironomus* larva. E. *Chironomus* eggs. F. *Chironomus* pupa.

Genus Spaniotoma (Orthocladius).—The eggs of this genus are laid in gelatinous strings. The larvæ are very similar to those of *Chironomus* but the body colour is usually yellowish, greenish, or bluish. The dorsal papillæ near the anus are short. They live in running water, either free, in a gelatinous case, or in tubes built of sand which are attached to stones (Fig. 154).

Genus Forcipomyia.—The long narrow worm-like body of the larvæ of this genus is very easily recognised. Except for the brownish head covered with hard skin it looks very like an aquatic worm in the resting position, but the swimming movements are different, the body being thrown into a series of stiff-jointed undulations passing from head to tail. There are no false legs on the body ; the last abdominal segment ends in a crown of stiff bristles within which are to be found two spiracles. The pupa is much shorter than the larva, but longer than those of related genera. The species winters as a larva, pupating and emerging in early summer. The larvæ are common among floating plants, particularly thread algæ, in still water (Fig. 153c).

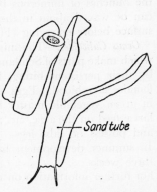

Sand tube

FIG. 154.—*Spaniotoma (Orthocladius).* Sand tubes of larva.

Genus Chironomus.—Species of *Chironomous* are the largest members of this family. The eggs are laid in gelatinous strings (Fig. 153E). Some species have red larvæ, the colour being due to a red pigment, hæmoglobin, in their blood ; in others the larvæ are colourless. The species possessing hæmoglobin usually live in mud tubes which they make out of mud particles and silk. Colourless species more generally live among floating vegetation. The tube dwellers hold on to the sides with their thoracic and abdominal pairs of false legs and undulate their bodies to keep up a current of water through their tubes so that they may obtain a supply of oxygen. They leave their tubes of their own accord for short periods, jerkily swimming to the surface, and if deprived of the tubes they build new ones. The second-last abdominal segment has four tubular "gills" on the ventral side, and the last segment has four retractile anal gills (Fig. 153D). The larvæ live on the organic material contained in mud, and they are found in all kinds of still water where there is a deposit of mud, sand, or organic débris on the bottom. Rain-water tubs and horse-troughs are sometimes full of them. The larva has very much swollen thoracic segments when it is ready to pupate. The pupa is usually formed in the mud tube, it is fairly active and normally protrudes its head from one end of the tube so that the bunches of numerous breathing filaments on the thorax can be waved about in the water. The pupa swims to the surface before emerging (Fig. 153F).

Genus Culicoides.—The minute midges, about $1\frac{1}{2}$ mm. long, which make parts of Scotland almost uninhabitable in summer by their persistent biting, belong to this group. Only the females take blood. Eggs are laid in masses of fifty or more in green vegetable matter beside running water. The tiny larvæ feed on the algæ, etc. ; they are smooth, without bristles and possess no false legs. The pupa is anchored to plants. In summer development from egg to adult may take only three weeks.

For further information on midges (Chironomidæ) see p. 215.

B. *BRACHYCERA*.—The following families have common genera which are aquatic during their larval life :—

Family Tabanidæ.   The Horse-Flies or Clegs

The adult flies are strongly built with very large eyes on the head and a broad fat thorax and abdomen. The females bite man voraciously. The larvæ of many species are either aquatic or amphibious. The eggs are laid on vegetation close to water or on objects projecting above the water-level. The larvæ have a minute head followed by three thoracic and eight abdominal segments, the last one having a breathing siphon. The head can be retracted into the first thoracic segment. The colour is usually pale green, cream, or brownish, and the body is rather soft and fleshy, often slightly dorso-ventrally flattened. The abdomen has a pair of false legs on each segment but these are very short (Fig. 155). The

two main tracheal tubes meet in a single opening at the tip
of the siphon. Before pupation the larva
leaves the water to pupate in the ground
two or three inches below the surface.
Larvæ of this family are commonly found
in streams or among damp water moss and
leafy liverworts.

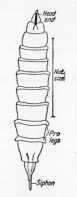

Genus Chrysops.—Full-grown larvæ usually
less than 2 cm. long, body striated.

Genus Hæmatopota.—Full-grown larvæ over
2 cm. long, surface of the body smooth.

Genus Tabanus.—Full-grown larvæ over
2 cm. long ; last abdominal segment
tapering with a fairly long extensible
siphon (Fig. 155).

FIG 155.
Larva of a Horse-Fly
(*Tabanus*).

ATHERICERA.—The two families with
common genera which are aquatic for
part of their lives are the Hover-flies (*Syrphidæ*) and the
Anthomyidæ.

Family Syrphidæ. Hover-Flies

The adults are fairly large and often brightly coloured ;
some species resemble bees or wasps so closely that you almost
have to count the wings to make sure that they have only one
pair and must therefore be flies. A number of species pass
their larval life in liquid mud or dirty water. One of the
commonest of these is *Eristalis*, the rat-tailed larva (Fig.
156A, B). The body is a dirty white, and the skin thick and
rough. The head is very small ; it is more or less retracted
into the thorax. The thoracic and abdominal segments are
wrinkled transversely but they are not sharply marked off
from one another. The thorax has one pair of short, blunt,
false legs and the abdomen six pairs ; the thorax bears in
addition a pair of short breathing trumpets. The last segment of the body is prolonged into a very long tail which
is in three segments which can be telescoped into one another.
The tip of the last segment is surrounded by eight feathery

hairs in the middle of which are the spiracular openings. The tail is extended until the tip is out of the water ; if the water becomes deeper the tail is further extended and so on until it reaches its limit. When fully extended it is several times the length of the body. The tracheal tubes leading from the spiracles are very large and easily seen through the transparent skin. The larva pupates inside the last larval skin

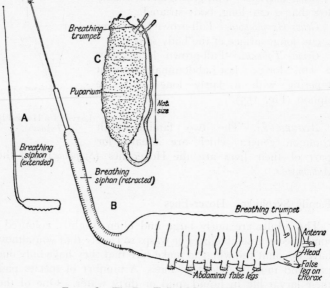

FIG. 156.—The Rat-Tailed Maggot (*Eristalis*).
A. Larva, natural size.　B. Larva, enlarged.　C. Pupa in puparium.

which retains the outward appearance of the larva, though the body is shortened (Fig. 156c). Short breathing trumpets are present on the thorax. The pupa in its puparium floats passively at the surface of the water.

Family Anthomyiidæ

This family of flies is closely related to the house-flies and bluebottles. Some have aquatic larvæ which are found

commonly among water-mosses. The form of the larva is
like that of an ordinary maggot with much tapered anterior
segments and no visible head, only
a pair of black mouth-hooks (Fig.
157A). The genus *Limnophora* has
pale yellow or whitish larvæ which
are common. They appear to
prefer situations where there is a
good current of water, as for instance among the plant growth
in a small waterfall. The larva
bears vestigial false legs on the
ventral side of each abdominal
segment, those on the last segment
being rather larger. There is a
pair of very small spiracles on the
prothorax and a much larger pair
at the hind end of the body on
the end of short tubes. The puparium (Fig. 157B) is reddish-
brown with eleven visible segments.

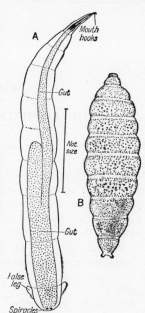

For the identification of aquatic
dipterous larvæ see

Aquatic Diptera, by O. A.
Johannsen.

FIG. 157.—*Limnophora.*
A. Larva.
B. Pupa. After Johannsen.

Part I. Nemocera, exclusive
of Chironomidæ and Cerato-
pogonidæ. 1934.

Part II. Orthorrhapha, Brachycera, and Cyclorrhapha.
1935.

Part III. Chironomidæ : sub-families Tanypodinæ,
Diameniæ and Orthocladiniæ. 1937. New York.

General

Biologie der Süsswasserinsekten, by von C. Wesenberg-
Lund (1943). Berlin, Wien.

Faune de France (1932), 23 Diptères Chironomidæ, par
M. Goetghebuer.

Handbooks for the identification of British Insects. Diptera 2,
Nematocera, by R. L. Coe, Paul Freeman and P. F.
Mattingly (1950). Royal Entomological Society,
London.

CHAPTER 10

THE ARTHROPODA : ARACHNIDS

THE Arachnids are a sub-division of the Arthropods and they include the Spiders, Scorpions, King Crabs, Ticks, and Mites. Only a relatively small number of Arachnids live in fresh water ; most members of the group live on land. In Britain there is one species of aquatic spider (*Argyroneta aquatica*) and over two hundred species of water-mites all of which belong to the *Hydracarina*. A group of microscopic creatures, the Bear Animalcules or *Tardigrada*, resemble the Arachnids in some respects and are usually placed among them. There is one species of Bear Animalcule which inhabits permanent ponds.

Arachnids possess the Arthropod characters of having many jointed legs and a hard chitinous cuticle (skin) covering the body, but they completely lack the sensory antennæ at the anterior end of the head which are so familiar in insects and Crustacea. The limbs on the head have various functions mostly concerned with feeding ; they are followed by four pairs of walking limbs. The segments of the body are fused together. In spiders the body is divided into two with a narrow " waist " between the divisions, but in the mites the body is in one piece. If you remember that the Arachnids have four pairs of walking legs then you will have no difficulty in recognising the fresh-water members of this group.

The Water-Spider.—Although a number of spiders live in damp situations near water, there is only one species which lives permanently below the water surface ; this is *Argyroneta aquatica*. In general appearance it is indistinguishable from many ordinary land spiders. The body is divided above into

two parts called the *prosoma* and the abdomen or *opisthosoma;*
the most anterior part of the prosoma is the head which bears

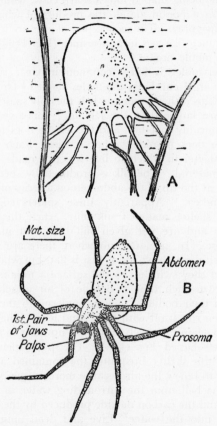

FIG. 158.—The Water-Spider (*Argyroneta aquatica*).
A. Submerged silken shelter. After Lulham.
B. Adult spider. After Brocher.

two pairs of jaws ; with this is fused the part of the body
corresponding to the thorax in insects to which are attached
four pairs of legs. The hind part of the body bears no legs and

shows little sign of segmentation. The jaws are not like those of the insects ; the first pair are made up of two joints only, the distal one being shaped into a sharp claw, while the basal joint contains a poison gland, the product of which is used to paralyse live prey. The second pair of jaws are long jointed palps very much like legs in appearance (Fig. 158B) and they function as tactile organs.

The water-spider is fairly common in clear still ponds. The females are 1 cm. long, the males about 1½ cm. long and have their palps rather swollen. It is the only spider in which the males are larger than the females ; moreover, the two sexes live amicably together, and the female does not eat the male as she does in many other species. Each constructs a separate shelter in which to live which is made out of silk and plant material. The silk is produced by special glands which open at the tip of the abdomen on a series of structures called spinnerets. Apparently these spiders construct an underwater shelter made of silk only, when they are kept in captivity and are not given suitable plant material for their purpose. In an aquarium where there are few plants the spiders spin a web which at first is flat, but when the web is completed they swim up to the surface a number of times bringing down each time a bubble of air attached to the hind end of their bodies ; they go underneath the web with this and then rub it off with their hind legs. The bubble of air floats up against the web pushing it upwards in the middle until it is bell-shaped (Fig. 158A). The spider spends most of its time in this shelter breathing the contained air, it does not require to renew the air because more oxygen will diffuse in under the bell from the surrounding water as the spider uses it up, and the carbon dioxide produced by the spider will diffuse out into the water. Live and dead animals which float on the surface of the water are captured by the water-spider ; these are carried down to the shelter and are consumed inside it. These spiders live very well in an aquarium as long as only one, or a male and female, are placed in the same container, otherwise they eat each other. One spider may construct a number of shelters. When swimming through the water the body is always covered by a coat of air which clings to the hairs on the body, this gives the spider a silvery appearance.

Air is breathed in through two small openings or stigmata on the ventral side of the abdomen. The first pair lead into a chamber in which are a number of flat plates like the leaves of a book, in each plate blood circulates and the structure is known as a "lung book." As well as these, bunches of tracheal tubes lead out from the gill chamber, but these do not branch as they do in insects. The second pair of stigmata communicate with the ends of tracheal tubes only.

The eggs are laid in summer in the upper part of a female's shelter which is partitioned off; she lays about thirty to seventy, each being about 1 mm. long. As a rule this shelter is made near the surface if eggs are going to be laid inside. The top portion is made of more dense silk than the lower part which the female continues to use as a shelter. When the young hatch they do not make shelters for themselves at once but they use instead a small empty snail shell which they fill with air.

The Water-Mites (Hydracarina)

The water-mites belong to the larger group of Acarina which contains the mites and ticks. Their general appearance is much like that of a small spider because they have a round or oval body with eight legs, but a close examination shows that their body is made of one piece, not divided into two, as it is in spiders. Most water-mites, if viewed from above, appear as brightly coloured round or oval-shaped animals with eight legs and one pair of palps, coming from underneath the body. The average size is about 2 mm., the largest British species being 8 mm. Somewhere near the anterior margin of the body are the eyes of which there are usually two pairs (see Fig. 165). Between the palps is a complicated structure called the *capitulum* or false head which is only visible in many forms when the animal is placed on its back. The capitulum encloses the mouth-parts which are adapted for piercing and sucking; it ends in a pointed snout called the rostrum, the actual mouth being at the end of this. When a mite is placed on its back and examined, it will be found that the legs articulate with flat plates on the ventral surface of the body; these plates are called *epimera* and they are important because their size and position helps in identifying a particular mite. Each leg has six joints which usually bear bristles; swimming hairs are also often present. The last joint has a pair of retractile claws. Either between or behind the bases of the epimera is a slit-like genital opening on either side of which is a semi-circular plate often bearing a number of cup-shaped structures of unknown function called *acetabula ;* these are also used

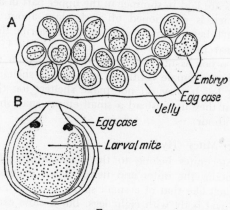

FIG. 159.
A. A group of eggs of a Water-Mite.
B. A single egg enlarged.

for purposes of identification. Near the posterior end of the body is the anus.

The life-histories of most water-mites are still incompletely known because most of them have a parasitic stage when they are young. If the eggs are obtained from an adult it is not possible to rear them unless you can provide the correct animal for the young to parasitise. The knowledge of what is the correct host depends on having previously reared the parasitic stage through to the adult. It is possible to get a fairly general idea of the diversity of habit by piecing together all the various information about the group. The females usually lay reddish coloured eggs which are enveloped in a hard transparent jelly covering (Fig. 159). In aquaria these are often deposited on the glass sides, but in nature the surfaces of leaves and stems of aquatic plants or submerged stones are used. The eggs may be laid singly or in groups. One genus (*Hydrarachna*) pierces tiny holes in water-plants and places its eggs inside. The eggs hatch in about a month or six weeks into a larva with only six legs and a prominent false head (see Fig. 160). The larva is practically always parasitic on some other animal, usually an

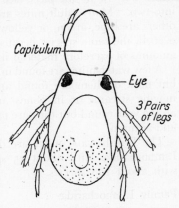

FIG. 160.

Larva of *Hydrarachna*, length 1 mm.

aquatic insect, but bivalve molluscs, fresh-water sponges, and even fish are known to be the hosts of some genera. Insects such as dragon-flies and pond skaters are fairly often found with small oval orange or red bodies attached to them ; these are larval stages of mites. In a few species the larva does not come out of the egg to find a host but remains inside passing the whole of the larval stage inside the egg, and in these cases the mite is never parasitic. After a period as a larva, mites usually enter a resting stage after which the skin is shed and the animal changes its form considerably (i.e. it metamorphoses), emerging from the larval skin with eight legs. The nymphal stage seems to be free living in most forms, the nymphs eating a carnivorous diet. Another period of rest followed by a second moult results in the nymph being transformed into an adult. All adult water-mites are carnivorous creatures ; they must not be overcrowded in captivity or they will eat each other ; they should be given suitable food, such as water-fleas or other small Crustacea. There is usually not much difference between the males and females, but the males may have slight modification of their third and fourth legs ; in one genus (*Arrhenurus*) the males all have the hind part of the body produced into a characteristic " tail " which may be in one piece (see Fig. 165B) or may be divided into two. Many adult mites are bright red or brownish-red, others are blue, green, or yellowish-green. The colour varies even in the same species because it is partly due to varying amounts of excretory material inside the body.

A few water-mites are found in streams with a strong current, but most of them are found among the vegetation of slow-moving streams or in ponds and lakes. The adults occur all the year round but they are more numerous in late summer and autumn.

The British water-mites are divided into two families, the Limnocharidæ with about 75 species and the Hygrobatidæ with about 128 species.

Family Limnocharidæ

The colour of the mites in this family is always some shade of red. Often they are quite large (5-8 mm. long), the skin is usually soft, sometimes with hard plates on the back. They have two pairs of eyes generally enclosed by capsules with in some species a " third eye " between the other pairs. On the ventral side the epimera of the legs do not occupy much space, they form four groups, the epimera of legs one and two and three and four being united on each side. The genital area is always near the epimera.

The seventy-five British species of this family belong to twelve different genera only a few of which are mentioned here.

Genus Protzia.—Eye capsules well separated. Hind pair of epimera well separated from front pair. Genital area between the front and hind pairs of epimera. No swimming hairs on legs. One species *P. eximia* found in mountain streams.

Genus Eylais.—Eyes in two capsules close together. No swimming hairs on fourth pair of legs. Genital area far forward with no acetabula. Twenty-three species found in slow-flowing water, marshes, or slightly salt-water.

Genus Limnochares.—Body rectangular in outline. Eyes

close together, short, stout, spiny legs. Hind pairs of epimera small, well separated from front pairs. One species *L. aquaticus*. This mite is found among mud at the bottom of lakes and canals, it does not swim at all but only creeps along slowly.

Genus Hydrarachna.—Body globular, often dark red, with the back well arched. Soft skin with usually one or more chitinised plates on the back. Eyes in pairs near mid-line with a single " third eye " between them. Legs short all except the first pair having swimming hairs. Genital area with

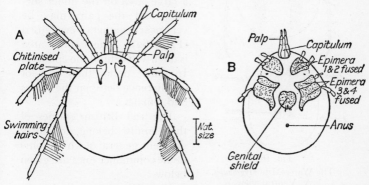

FIG. 161.—The Water-Mite *Hydrarachna*. After Soar and Williamson.
A. Dorsal view. B. Ventral view.

numerous small acetabula on the plates which are fused to form a shield covering the genital area. Twenty-two British species (Fig. 161). The females pierce the stems of water-plants and place their eggs inside. From the eggs hatch rather peculiar larvæ with the head end of the body much enlarged (Fig. 160). These swim about actively in search of some insect host to parasitise, usually a water-bug. They attach themselves to the skin of the bugs and change into a red shapeless oval mass. *Nepa* (p. 144) is quite often found with several dozen of these mites. After feeding on the insect for some weeks the mites shed their skins and leave their host to become free living.

Family Hygrobatidæ

The colour of the mites in this family is very varied ; a few are red, but brown, yellow-green, and blue or mixtures of these colours are more common. The skin of some is soft while others have hard skins (heavily sclerotised). The two pairs of eyes are not enclosed in capsules and there is never a median eye. The epimera often cover a large area on the under surface, and the pairs may be very near together or even fused (Fig. 162). The genital area is often well behind the epimera. In some species there is a lot of difference between the males and females.

The 128 British species of this family belong to thirty different genera, of which some of the commoner ones are given below.

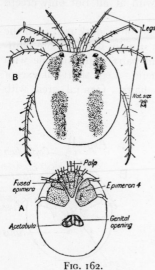

FIG. 162.
The Water-Mite *Hygrobates*.
A. Ventral view.
B Dorsal view.

Genus Hygrobates.—Body soft-skinned, oval in shape. First pair of epimera fused with the capitulum (see Fig. 162). No swimming hairs on the legs but plenty of bristles ; the plates on either side of the genital opening with three or more pairs of acetabula. Members of the genus are very common in weedy ponds particularly among duckweed. Colour usually yellowish-green or blue with dark markings on the back.

Genus Megapus.—Small mites generally less than 1 mm. long. Round in outline or a long oval. The first pair of legs as a rule longer than the second. The epimera fused together (see Fig. 163) ; genital area has three acetabula. There are five species found among floating water-weed. The nymph has a number of long hairs at the hind end of the body (see Fig. 163B). This species is very sluggish.

Genus Unionicola.—Body soft-skinned, oval in shape,

broadening at the back. First pair of legs usually stouter than the others. Last pair of epimera very large and rectangular. Genital area near posterior end of body with ten or more acetabula. Seven British species. The larvæ of one species, *U. Crassipes*, have been found as parasites of the fresh-water sponge *Spongilla ;* another *U. bonzi* is found in the adult stage inside fresh-water mussels. It is reasonably common, so if you are ever dissecting *Anodonta* (see p. 244) you should look out for it.

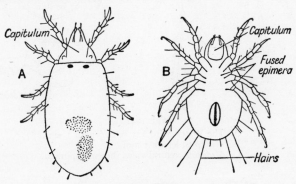

FIG. 163.
A. *Megapus* adult, length 1 mm.
B. *Megapus* young mite, ventral view.

Genus Neumania.—Body soft, oval in shape. Usually a more or less distinct Y-shaped reddish area on the back (see Fig. 164A). Palps less stout than the first pair of legs. The third pair of epimera distinctly marked off from the fourth pair. On the fifth segment of the fourth pair of legs are a number of specially long bristles. Rounded genital plates with numerous small acetabula. There are five British species, some of which are common.

Genus Mediopsis.—In this genus the body is flattened on top and almost perfectly circular in outline (see Fig. 164B) or oval. The eyes are on the extreme edge of the anterior part of the body. The epimera are united at the edge into one group which covers a large area of the ventral surface. The genital area is far back with three acetabula on each genital plate. There are two species of which *M. orbicularis* is common.

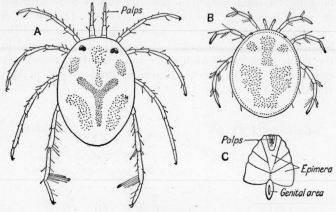

FIG. 164.

A. The Water-Mite *Neumania*, length 1 mm.
B. *Mediopsis*, length 1 mm.
C. *Mediopsis*, ventral view, showing epimera.

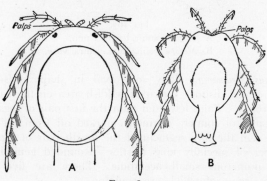

FIG. 165.

A. *Arrhenurus*, female.　　B. *Arrhenurus*, male.

Genus Arrhenurus.—Body covered by a thick skin, shape oval in female, the male possesses a " tail-piece " (see Fig. 165B). Palps short and stout. The first pair of epimera are joined behind the capitulum. The second, third, and fourth pairs of legs have swimming hairs. In the female the genital area lies just behind the epimera, in the male it lies at the junction of the " tail-piece " with the body. There are twenty British species most of them green or brown in colour.

Book for further reference—
Soar C. D., and Williamson, W., British Hydracarina. The Ray Society, London, Vols. 1-3 (1925, 1927, 1929).

Water-Bears or Bear Animalcules (Tardigrada).— When you are examining samples of sediment from the bottom of a pond or a rain barrel under a microscope, you may come across a " water-bear " amongst the other microscopic creatures, such as Protozoa and Rotifers, which you find in such places. Most species of water-bears are land animals living in damp places, but the species *Macrobiotus macronyx* lives in fresh water (Fig. 166). When highly magnified these tiny creatures have a superficial resemblance to a bear. However, they possess eight short fat legs ending in claws and they are covered by a transparent flexible skin. The head end is well defined

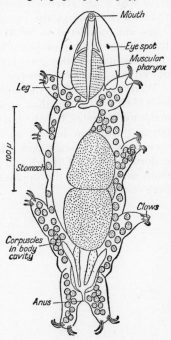

FIG. 166.—The Water-Bear
(*Macrobiotus macronyx*).

and carries a pair of small eye-spots. The mouth is at the extreme tip of the head ; it leads past some stylet structures into a very muscular pharynx with which these animals suck plant juices. Leading out from the pharynx is a narrow part of the gut which connects with the wide sac called the stomach ;

this narrows at the hind end giving off a pair of lateral pouches before it connects with the anus situated between the last pair of legs. Between the internal organs and the thin body wall is a large space filled with fluid in which conspicuous round cells (corpuscles) float about. In the head region is a pair of salivary glands, and in the middle of the body a mass of tissue producing eggs is usually present. They have a well-developed muscle system as well as a small " brain " from which are given off nerves so that for their size they rival the Rotifers in the complex nature of their structure. They are considered to be Arthropods because of their four pairs of legs and because they have a chitinised skin, but since they resemble no other group of these animals closely, they are placed in a class by themselves.

Water-bears feed on the sap of plants by piercing the cells with their stylets and then sucking out the juice. There are males and females, but the latter are the more numerous and larger. The skin is shed at intervals and the females lay their eggs in the cast skin. Land-dwelling species go into a dormant state if dried up, recovering when placed in a moist situation, but this apparently does not happen with the fresh-water species.

CHAPTER 11

THE MOLLUSCA

FRESH-WATER SNAILS, LIMPETS, AND MUSSELS

THE Fresh-water Snails, Limpets, and Mussels belong to the Phylum Mollusca, a large group of animals having soft bodies without any sign of segmentation, a fold of the body wall (mantle) outside which is usually a hard external shell. Inside this the whole animal can be withdrawn. Most of the fresh-water forms are reasonably large so that their habits can be studied in a laboratory or classroom without the aid of a microscope ; many species live well in an aquarium.

Water-snails are similar in appearance to the familiar land ones. Their shells are made of one piece (hence the name Univalve molluscs) arranged in a number of whorls which get bigger as they get nearer the opening through which the snail protrudes its body. When the snail hatches from its egg it occupies the smallest whorl only, and as the animal grows it adds more and successively larger whorls on to the original one, so that the size of the shell keeps pace with the increasing size of the body. The substance of the shell is produced (secreted) by cells round the edge of the mantle, the part of the snail which is immediately beneath the shell covering the internal organs. Some snails have a horny plate—the *operculum*—attached to the foot ; this plate fits over and closes the opening to the shell when the animal has withdrawn itself. These snails are called *Operculates*, and there are ten fresh-water species in Britain. The *Operculates* breathe the dissolved oxygen of the surrounding water through special gills, and are chiefly found in running water which is reasonably aerated. The rest of the water snails are without an operculum, and they may come to the surface to breathe air taken into a chamber under the mantle which acts as a lung ; these are called *Pulmonates* and there are twenty-six British species. Fresh-water limpets are really snails (of the pulmonate type) in which the whorls of the shell are not evident (see Fig. 172).

The fresh-water mussels and some much smaller related molluscs, sometimes called fresh-water cockles, have shells made up of two separate pieces or valves which are hinged together along the back ; they are known as *Bivalve molluscs*. As in the case of the univalves (snails) the shell of a bivalve is small when the animal is young, but additions are made to it by the edge of the mantle and these additions are visible as ridges called lines of growth (see Figs. 178 and 182). By counting them you can get some idea of the age of your specimen. The internal structure and the mode of life of these two mollusc groups (univalves and bivalves) are very different, and for that reason I propose to deal with them separately.

(a) Snails ; Univalve Molluscs (Gastropoda)

The snails and limpets are fairly active, gliding about by muscular and ciliary action of a flat foot on the surface of stones and plants and feeding principally on the green algal slime which covers them. When they are gliding the head is thrust out from under the shell ; there is a pair of tentacles on the head with eyes at or near their bases. The mouth is at the extreme front. Running backwards from the head is the main part of the body, the foot, on which the animal glides ; it is muscular, and is kept covered with slime produced in a gland which opens close behind the mouth. The slime helps the animal to glide easily. All the principal organs, such as the heart, digestive system, and egg and sperm-producing tissues (gonads) are in that part of the body which is tucked away inside the shell and can only be seen after its removal. In very young snails, however, some of the internal structures can be seen through the shell, which is more transparent at this stage. Just before hatching, or soon after, the heart may be seen beating if the specimen is placed under a dissecting microscope.

When feeding, the food is rasped off by the *radula* which is an apparatus inside the mouth consisting of rows of teeth arranged transversely across a movable band which is worked backwards and forwards against surfaces. As the teeth wear out they are replaced from behind by the band growing forward. If you place some common pond snails (such as the wandering snail, *Limnæa pereger*, or the ear-shaped snail, *Limnæa auricularia*) in a glass aquarium tank which has contained water for some time and has a good green growth of algæ on the sides, you may see the snails gliding slowly along leaving a narrow trail of clean glass where they have eaten away the algæ. You may see the snail's mouth open and shut, and with a hand lens you will be able to make out the radula working.

Those snails which have no operculum (and therefore cannot close the opening to their shells) normally come to the surface of the water at intervals to breathe. They come partly out of the water, and take some air into a cavity under the shell. This cavity is lined above by the thin mantle which has a rich blood-supply, and below by the main body of the snail (see Fig. 177B). The mantle acts as a lung and takes in oxygen from the air. As well as this method of obtaining oxygen, any tissue in direct contact with the water, such as the head or foot, is probably able to take up some of this gas if there is any in solution in the surrounding water (many pulmonates live in stagnant water containing little or no dissolved oxygen, when they must perforce breathe exclusively with their lungs). In fact the small fresh-water limpets are able to get sufficient oxygen through their body-surface from solution, while larger snails come up to breathe more frequently the less oxygen the water contains. If two similar snails are placed, one in well-aerated water and the other in water from which all oxygen has been driven by boiling (after boiling it must be cooled to the same temperature as the aerated water), the snail in the previously boiled water will be seen to come up to breathe much more often than the other.

The breeding habits of snails are various. All the

pulmonate types have male and female organs together in the same animal (hermaphrodite). They all lay eggs which seem to be normally cross-fertilised, but should this be impossible then they are able to fertilise their own eggs. The eggs are laid in clear, transparent, gelatinous capsules attached to pond weeds, stones, or floating objects. In aquaria they are often deposited on the glass sides. Large pond snails may lay as many as 500 eggs. Most operculate snails also lay eggs, but two species are viviparous producing well-developed young. One of these, the large fresh-water winkle (*Viviparus viviparus*), (see Fig. 167) has individuals which are either male or female. The other (*Hydrobia (Paludestrina)*

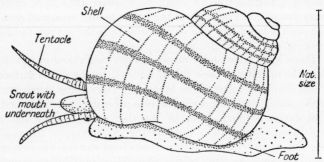

FIG. 167.—The Fresh-Water Winkle (*Viviparus viviparus*)

jenkinsi), (see Fig. 169) produces its young without fertilisation (parthenogenetically), and no male specimens have ever been found.

The following are the snails and limpets found in fresh water in Britain :—

The univalve fresh-water molluscs all have a shell made of one piece usually arranged in whorls. Most resemble a garden snail in general appearance. They are divided into *Operculates* which have an operculum with which they can close the entrance to their shells, and the *Pulmonates* which have no such horny plate and which usually breathe air directly.

Operculate Snails.—These possess gills (see Fig. 170D) and are usually found in running water.

The Fresh-water Winkles, Genus Viviparus.—Two species, *V. viviparus* (Fig. 167) and *V. fasciatus* are large snails found in slow-running water and at the edges of large rivers in the English counties as far north as Yorkshire. They are viviparous (producing fully developed young instead of laying eggs) and have about fifty young at a time ; the sexes are separate, each snail being either a male or a female. *V. viviparus* may be so common that the rocks at the edge of a river (e.g. the Trent in Notts.) may be almost covered with specimens of all sizes. This species has a brownish-yellow shell with three very distinct dark bands running round each whorl ; the operculum is thick and conspicuous on the dorsal part of the foot. The tentacles are well developed and afford a means of distinguishing the sexes, the male's right tentacle

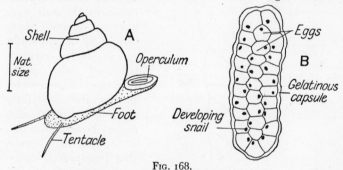

FIG. 168.
A. *Bithynia tentaculata.* After Lulham.
B. Egg mass enlarged. After Wesenberg-Lund.

being shorter and stouter than the female's. *V. fasciatus* is less common and more usually found in the south-eastern counties. It has a thinner operculum and the shell has a more pointed spire than that of *V. viviparus.*

Genus Bithynia.—There are two species of this genus, *B. tentaculata* (Fig. 168) and *B. leachii*, of which the first is much the more common, occurring throughout England, Ireland, and Scotland as far north as Edinburgh. Both are found in slow-flowing water, such as canals, ditches, rivers, and occasionally in ponds. *B. leachii* is found in the south and Midlands of England. These snails are small, only measuring $1\frac{1}{2}$ cm. long when full grown, with very thin tentacles (which are similar in males and females) and having a hard brittle operculum. The eggs are laid in a gelatinous capsule, containing about three rows of eggs, which is attached to water-plants or stones. Full-grown specimens of *B. tentaculata* have six whorls to their shell and measure about 1½ cm. long, while *B. leachii* is smaller with only four or five whorls, and the shell has a much more pointed apex.

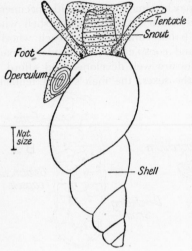

FIG. 169.—Jenkins' Spire Shell (*Hydrobia* (*Paludestrina*) *jenkinsi*)

Genus Hydrobia (Paludestrina).—Only one species, Jenkins's Spire Shell (*H. jenkinsi*, Fig. 169), a small blackish or brownish snail with a turret-shaped shell. Originally a salt-water species, it has recently colonised and spread over most of the country during the last forty years. In canals (such as the Chesterfield canal) it is now so common that hundreds of specimens may be taken with one scoop of a net. The eggs develop inside the female's shell without fertilisation and broods of twenty to thirty fully developed young are produced. No males have ever been found. This means that if one snail survives transportation to a new ditch or canal it may colonise it. Apparently this species is free from any parasitic flatworms (see p. 38), so that its numbers are not kept down by such natural means.

Genus Valvata.—Three species all easily distinguished from one another by the shape and size of their shells. These snails have the anterior end of the head prolonged into a very definite snout. Their long tentacles have eyes at their bases, and the anterior part of the foot is divided into two triangular portions (Fig. 170D). On the right side, coming from under the shell, two structures, a feathery gill and a long branchial filament, may be seen. The operculum is conspicuous. The commonest species is the valve snail (*V. piscinalis*), which is found all over Britain and Ireland in running water or at the edges of lakes. The shell is composed of five or six whorls arranged in a definite spire of about

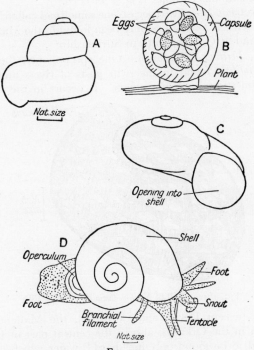

FIG. 170.

A. Shell of Valve Snail (*Valvata piscinalis*).
B. Egg capsule of the Valve Snail.
C. Shell of *Valvata macrostoma*.
D. *Valvata macrostoma*.

9 mm. in height (Fig. 170A). *Valvata macrostoma* is fairly common in the south, east, and Midlands of England. The shell is only 2 mm. high and about 4 mm. broad, being coiled in a fairly flat spiral (Fig. 170C). *Valvata cristata* is found among dense plant growth in streams and ditches all over the British

Isles, but is not very common. The shell is just about a millimetre high and 2-4 mm. broad, the spiral in this case being absolutely flat as in a small trumpet snail (see p. 242), from which, however, it can be easily distinguished by the presence of an operculum. All three species lay their eggs in flat gelatinous capsules (Fig. 170B).

Genus Amnicola.—Two species, one sometimes called Taylor's Spire Shell (*A. taylori*) has only been found in the Manchester canals and at Grangemouth in Stirlingshire.

Genus Neritina (Theodoxus).—One species only, the Nerite (*N. fluviatilis*), found in the swift parts of rivers and canals sitting on stones or water-plants. It occurs in the southern

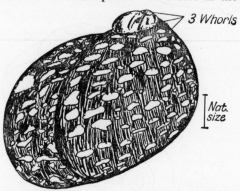

3 Whorls

Nat. size

FIG. 171.—Shell of *Neritina fluviatilis.* After Ellis.

counties (except Devon and Cornwall), the Midlands, and one loch in the Orkneys, also in the central part of Ireland. It is said to be found only in " hard " water which contains lime salts. The shell (Fig. 171) may be yellow, brown, or even black ; it is mottled with white, pink, or purple variega-tions. It consists of three whorls only, forming a low spire. The largest whorl has a big aperture which can be closed by a thick reddish-coloured operculum.

Pulmonate Snails.—These do not possess an operculum and they breathe through the inner wall of the mantle cavity which is vascular and acts as a lung. Breathing air they are therefore able to live in still or stagnant water.

The Fresh-water Limpets. Genus Ancylastrum and *Ancylus*.—The two species belonging to this genus are the River and Lake Limpet. They have thin fragile shells which are rather hook-shaped in side view (Fig. 172A). When the animals are moving, very little of their bodies appears from under their shell (see Fig. 172B). To see their structure they should be placed on a microscope slide with a little water, wait for them to grip the glass, then turn the slide upside down over a dish so that the limpet is covered with water. You will then be able to see them

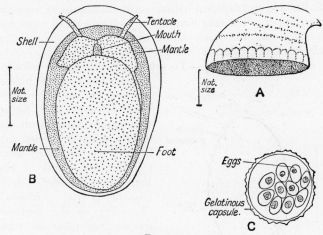

FIG. 172.

A. Shell of the River Limpet *Ancylastrum (Ancylus) fluviatile.*
B. The River Limpet, viewed from below.
C. Egg capsule of the Lake Limpet (*Ancylus lacustris*).

moving from underneath, when the foot, mouth, and tentacles will be readily made out. The River Limpet (*Ancylastrum fluviatile*) clings to stones in the beds of swift rivers and streams, and at the edges of lakes where there is some wind action. It appears to need a clean situation and cannot stand getting silt round its shell, probably because it is then unable to breathe. When disturbed it clings tightly to its place of attachment and has to be removed with care because the shell is easily broken. The edge of the shell is soft and thus fits into any irregularities in the surface so that the shell can be held down firmly.

The eggs, two to ten in number, are laid in a round transparent capsule, which is attached to stones. The river limpet is found everywhere in Britain and is definitely common, but it does not live well in a non-aerated aquarium. The Lake Limpet (*Ancylus lacustris*) is smaller, being seldom more than half a centimetre long. It is found in lakes, ponds, and in slow-flowing water at the sides of rivers. It does not occur in the north of Scotland or many parts of Ireland but it is common in the south of England and the Midlands. The spire of the shell is twisted to the left ; this, together with its smaller size and different habitat easily distinguish it from the river limpet. The lake limpet lives well in a small aquarium among stones and water-plants where it readily lays its flat gelatinous egg capsules (see Fig. 172c).

Genus Aplecta (*Physa*).—To this genus belong the so-called Bladder Snails of which there are five species. They have very polished pale brown transparent shells with the whorls going to the left side instead of the more usual right-handed spiral. The Fountain Bladder Snail (*Aplecta* (*Physa*) *fontinalis*), (Fig. 173) is the common species, occurring in most parts of Britain where there is clean running water containing plants, as for instance in a water-cress bed. In this snail part of the mangle is reflexed over the shell in a number of tongue-like processes (see Fig. 173) and the shell itself is very fragile. It will live and breed in a large aquarium but it cannot survive in stagnant water. The Moss Bladder Snail (*Physa* (*Aplecta*) *hypnorum*) is much rarer. The shell is larger, more pointed, and there are no processes of the mantle reflexed over the shell. One other species, *P. acuta*, is found in artificial tanks and canals, such

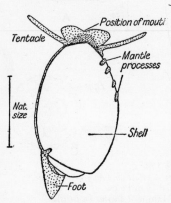

Fig. 173.—The Fountain Bladder Snail (*Aplecta* (*Physa*) *fontinalis*).

Position of mouth

Tentacle

Mantle processes

Nat. size

Shell

Foot

as those of Manchester, Birmingham, Chester, Shrewsbury, Aylesbury, and Cardiff. The eggs of the bladder snails are

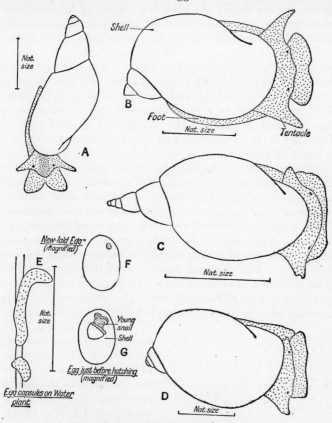

FIG. 174.—Pond Snails : Genus *Limnæa*.

A. The Dwarf Pond Snail (*L. truncatula*).
B. The Ear-shaped Pond Snail (*L. auricularia*).
C. The Giant Pond Snail (*L. stagnalis*).
D. The Wandering Snail (*L. pereger*).
E, F, G. Eggs of the Wandering Snail (*L. pereger*).

laid in the usual gelatinous capsules and the young of the Fountain Bladder Snail (*A. fontinalis*) have very pronounced "tails" coming out from the mantle.

Pond Snails. Genus Limnæa.—This is a genus containing a number of very common large pond snails all of which have brown conical shells ; the head bears a pair of flat triangular tentacles which are not retractile like those of some other snails. Each tentacle has an eye at its base. The largest is the common pond snail (*Limnæa stagnalis*) which has a conical shell up to 5 cm. long (see Fig. 174c). It is found in ponds, preferably large ones, also lakes, canals, and slow rivers in most of England, Ireland, and a few places in south Scotland. The eggs are laid in a large sausage-shaped gelatinous capsule attached to plants. The Wandering Snail (*L. pereger*, Fig. 174D) is the commonest of all British fresh-water snails, and is found all over Britain in ponds, lakes, ditches, marshes, and in running water. The shell is conical with the last whorl very large, as is also the opening into the shell. If these snails are kept in an aquarium during spring and early summer, numerous long gelatinous egg masses will be deposited. These should be inspected every few days to see how the eggs develop into tiny snails. Just before the eggs hatch the young snails will be found gliding round the inside of their egg (see Fig. 174G). This of course applies to the development of all pond snail eggs, but those of the wandering snail are the easiest to obtain. The Ear-shaped Snail (*L. auricularia*) is fairly common in large areas of water, chiefly in the south and east of England, though it is found in Scotland and Ireland as well. The last whorl of the shell is again large and the opening into the shell is enormous, and typically it is ear-shaped (see Fig. 175A) ; the whole shell has a very squat appearance with a pointed spire. Sometimes this snail breeds before the shell has reached its full

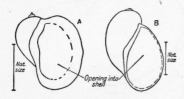

FIG. 175.—The shell of
A. Ear-shaped Snail (*Limnæa auricularia*)
   contrasted with
B. The Wandering Snail *Limnæa*
   pereger).

development, and in this case the characteristically much-expanded mouth to the shell is not present and the animal is very like the wandering snail, but the shell spire is more pointed. Breeding takes place early in the year and after that the adults die off, so that it is often not possible to find any adults after June. The Bog Snail (*L. palustris*), (Fig. 176B) is a much smaller species with a thick conical shell which has rather a narrow opening ; it lives in varied places, including peat bogs, but is not very common. It lays between sixty and eighty eggs in a cylindrical capsule, and has the habit of crawling out of any aquarium that is not covered. Two more species are found in marshy grass or in ponds which frequently dry up. One, *Limnæa*

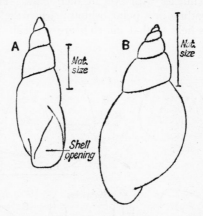

FIG. 176.—Shells of

A. *Limnæa glabra.*

B. Bog Snail (*Limnæa palustris*). After Ellis.

glabra (Fig. 176A), is found in temporary ponds in meadows where it is sometimes the only snail, while the other, the Dwarfed Limnæa (*L. truncatula*), is found on dry land near water as well as in it. This snail (see Fig. 174A) is the usual host for part of the life of the parasitic Liver Fluke which attacks sheep (see p. 37). Sheep only suffer from this parasite when they are kept on land which is damp and not properly drained, where this snail can exist.

Genus Myxas (*Amphipeplea*).—The only species, *M. glutinosa*, is rare, and the shell is difficult to distinguish from that of the Wandering Snail (*Limnæa pereger*, see p. 240 above), but it is thinner and more glossy in appearance. The live snail has an exceptionally large foot.

Ramshorn Snails. Genus Planorbis.—All the snails of this genus (of which there are fourteen British species) have their shells coiled in a flat spiral. Another peculiarity is that their blood is coloured red by the pigment hæmoglobin ; this no doubt makes their blood a more efficient oxygen carrier so that these snails may possibly be more fitted to live in stagnant water. All the species have long thin non-retractile tentacles with eyes at their base. The foot is rather short, ending in a point behind ; when the animals glide, the shell is held so that the coils are vertical. Not all the species are

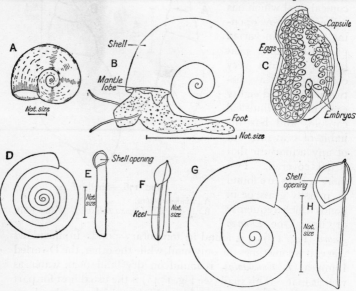

FIG. 177.—Ramshorn (Trumpet) Snails : Genus *Planorbis*.

A. Shell of *Planorbis albus*.

B. The Common Trumpet Snail or Ram's Horn (*Planorbis corneus*). After Lulham.

C. Egg capsule of the Common Trumpet Snail (*P. corneus*). After Wesenberg-Lund.

D and E. Shell of the Round-spired Trumpet Snail (*P. spirorbis*).

F. Shell of the Keeled Trumpet Snail (*P. carinatus*).

G and H. *Planorbis complanatus* shell from two angles.

easily distinguished ; attention should be paid to the size of the shell, the number of whorls, the flatness of the coil, and the appearance of the shell entrance.

The Ram's Horn shell (*Planorbis corneus*) is much the largest species (see Fig. 177B). It has a dark reddish-brown shell of

2½ cm. in diameter, and it is not likely to be confused with any of the others. It occurs chiefly in the Midlands and south-eastern England, in ponds, lakes, and slow rivers ; in some places it is very common. About sixty eggs are laid at a time in a flat gelatinous capsule which is generally fixed to a plant (Fig. 177C). This species can be obtained from aquarium dealers ; sometimes an albino variety is supplied which lacks the brown pigment in its body and shell with the result that the red blood, showing through by transparency, makes the animal crimson coloured.

There are three medium-sized species which are often difficult to separate. These are the Flat Ram's Horn (*Planorbis complanatus*), the Keeled Trumpet Snail (*P. carinatus*), and *P. acronicus*. The last-mentioned is a rare species found only in the Thames valley. Of the other two, the keeled trumpet snail can be distinguished from the flat ram's horn by a central keel (see Figs. 177F, H). Both species are found together in slow-running water, ponds, and marshes in most of England, Ireland, and south Scotland.

Another group of three very similar species is the Whirlpool Trumpet Snail (*P. vortex*), the Round Spired Trumpet Snail (*P. spirorbis*), and a rare species *P. vorticulus*, which is only found in Norfolk and Sussex. All these snails have very flat-coiled shells made up of a large number of whorls. The first two are definitely common, often being found together in ponds, ditches, rivers, and lakes throughout the country. The whirlpool trumpet snail may be distinguished from the other by its even flatter coil and the presence of a keel towards the top of the whorl ; it tends also to have fewer whorls than the round-spired trumpet snail (see Figs. 177D, E).

Two small species with shells, of a few coils rapidly increasing in width are *P. albus* and *P. lævis*. Both are widely distributed, *P. albus* being common in ponds, even very small ones, over most of Britain. A clean shell of *P. albus* is whitish-grey with definite striations (see Fig. 177A), while that of *P. lævis* is smooth ; the latter snail is rare.

The Twisted Ramshorn Snail (*P. contortus*) is a small species common in running water ; the whorls of the shell are more tightly compressed than those of any other *Planorbis* ; one

side of the shell is concave while the opening into the shell
is crescent-shaped ; size about 5 mm. in diameter. *P. nautileus*
can be recognised by its very small size, it is only about 2 mm.
in diameter.   It is found among weeds in stagnant or running
water.   One other species is widely distributed, *P. fontanus*,
which has a very highly polished shell, shaped like a quoit.
The last species to be mentioned is *P. dilatatus* which is only
found in the Manchester canals and more recently in the
Hebrides.   This small snail has a shell about 3 mm. in
diameter with a large opening which is not quite in the same
plane as the other coils of the shell.

Genus Segmentina.—One rare species, *S. nitida*, is found in
marshy districts in south-east England, the Severn valley,
and in several places in Yorkshire.

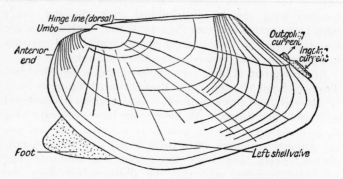

FIG. 178.—The Swan Mussel (*Anodonta cygnea*), natural size.

(b) Mussels and Cockles ; Fresh-water Bivalves (Lamellibranchia)

The fresh-water Bivalve Molluscs are represented in Britain
by a number of large fresh-water Mussels and by the much
smaller and more numerous " Orb-shell " and " Pea-shell "
Cockles.   As already stated the internal structure and way
of living of the bivalves differs greatly from that of the snails.
They are much less active and spend most of their time partly
embedded in mud or fine shingle at the bottom of a pond or
lake, river, or canal.   For this reason they often escape notice,
though common enough.   In some places if you go for a swim

in a lake there is a zone near the edge of what feels to your
feet like flat stones with sharp edges ; these are actually large
mussels partly embedded, and the sharp parts are the edges of
their shells ! The structure of a bivalve is best understood
by examining a large mussel such as the Swan Mussel *
(*Anodonta*), (Fig. 178). The shell of a full-grown specimen
is 10-15 cm. long, the two halves hinged together down
the back. A small humped area near the hinge is the
umbo, which is the oldest part of the shell. Round the
umbo the shell has increased in size in a rather uneven manner,
so that less material has been added to the blunt end of the
shell (which is the head end) than to the pointed end, which is

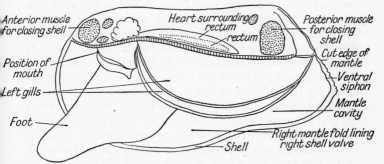

FIG. 179.—The Swan Mussel (*Anodonta cygnea*). Structures seen after
removal of left shell valve and mantle fold.

posterior. When the animal is alive and undisturbed the
two valves of the shell are slightly open all round except
down the hinge line ; the anterior two-thirds of the shell lie
buried in soft mud. At the posterior end are two short funnel-
shaped structures, the siphons, while near the head end a
muscular triangular piece of the foot may be thrust out ;
this is used for burrowing. No more of the animal ever shows
between the shell valves, so in order to see more, one valve
must be carefully removed. When disturbed or handled a
live mussel shuts the two valves of its shell tightly together
and holds them shut by means of very powerful muscles. To
see all the internal features the shell must be prised open,

* Description of the anatomy may be found in many text-books.

one half cut away from the muscles and removed. Lying immediately under the shell is a layer of tissue called the mantle ; it must also be carefully cut away from its line of attachment to the body, which is about 2 cm. below the hinge. The organs of one side of the animal will then show as in Fig. 179. On the other side is an exactly similar arrangement with the large median foot between. The mouth is difficult to find ; it is at the base of the foot near the anterior end. A pair of yellowish corrugated flaps are the gills, and dorsal to them at each end of the animal may be seen the muscles which held the shell together. In the mid-line near the hinge margin is a transparent patch of tissue through which the heart can be seen to beat if the animal is still alive (the rate of beat is slow being only about 4-6 beats per minute). The whole of the gills and a pair of triangular flaps (*palps*) round the mouth are covered by cilia which beat in such a way as to bring a current of water into the shell through the lower of the two siphons ; this current passes over the gills and *through* them, for these structures are not solid but are like a complicated gridiron with bars separated by perforations. Above the gills is a cavity communicating with the upper siphon (see Fig. 179) through which a current of water goes out from the shell. This current is very important to the animal because it brings it its food in the form of small particles in suspension, as well as a supply of oxygen which is absorbed by the gills and the mantle. A surprisingly large volume of water passes through these mussels in a day. It has been calculated that a sea mussel 3 in. long uses 15 gallons ! The gills are covered by a sticky mucilaginous secretion in which any food particles become entangled. The direction of the beat of the cilia drives the entangled food particles along the edge of the gills towards the mouth, where it is picked up by the cilia on the palps round the mouth, which therefore receives a constant ribbon of mucilage containing food, so long as the animal maintains a current of water through its siphons. This method of feeding, filter feeding as it is sometimes called, is in accord with the sedentary life of these animals. A small piece of gill from a live mussel shows very well how cilia beat. If a piece of gill is removed and placed on a slide with a little water, and examined under the low power of a microscope, the action of the cilia can be made out ; those near the edge of the gill are especially large and may be seen pushing food particles along. The mouth leads into the gut which is a long tube ciliated internally, coiling round and round in the interior of the foot ; ultimately it passes upwards and goes through the heart to the anus which lies near the dorsal siphon. The structure of all the bivalve molluscs is on the same general plan, though their breeding habits are rather different and will be considered individually.

First of all we shall consider the Mussels of which there are

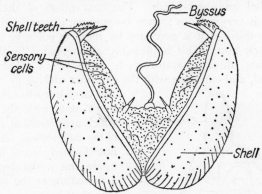

FIG. 180.—Glochidium larva of the Swan Mussel (*Anodonta*) as cast out from the parent shell. After Latter.

seven species. Two species belong to the genus *Anodonta*, three to the genus *Unio*, one to *Pseudanodonta*, and one to *Dreissensia*.

Genus Anodonta.—This includes the Swan Mussel (*Anodonta cygnea*) and *A. anatina*, both large molluscs living in firm mud at the bottom of canals, slow rivers, and some lakes. They occur in most parts of England, only sporadically in Scotland, and the Swan Mussel is the only one found in Ireland. Individuals are usually male or female, but hermaphrodites apparently occur. The eggs are produced during June-August in huge numbers—perhaps as many as half a million, and these are retained in the outer gills which form brood pouches; here they are fertilised by sperms from a nearby male coming in with the feeding current. They remain in the brood pouches for about nine months, during which time they develop into a special larva called a *Glochidium* (Fig. 180). In spring large mussels are often full of small yellow hard objects, the glochidia, which possess strong shells. If a few are placed under a microscope while still alive, they will be seen to open and shut their shells rapidly. Between the shell valves emerges a long sticky thread, the *byssus* (Fig. 180). At this stage they escape from their mother's protection into the surrounding water where their long sticky byssi may become entangled on water-weeds. The next phase of their life-history is very curious; they become parasites on fish. To achieve this end it is necessary for them to come in contact with their host. This is probably accomplished by the fish swimming through water-weeds to which glochidia are attached, and emerging with the glochidia sticking to its skin.

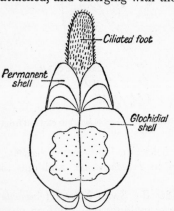

Permanent shell — Ciliated foot — Glochidial shell

FIG. 181.—Young Mussel shortly after leaving fish host. After Latter.

The larva burrows under the host's skin with the teeth which are present at the tips of the shell valves. A cyst is now formed round it, and inside this it develops into a miniature of the adult, a new shell being secreted on the inside of the larval one. This takes about three months, during which time all the food is obtained from the surrounding host tissues; finally the tiny mussel drops off and leads a free existence on the bottom (Fig. 181). In its parasitic stage the mussel may be carried far afield by the fish, which aids the dispersal of the species, though of course many young mussels must perish because they find the place where they leave the fish unsuitable for their further development (i.e. they may drop on to a stony river bed instead of on to firm mud which is essential for their mode of life). Each year the size of the shell is increased by several " growth rings " (see Fig. 182), apparently about three rings represent one year's growth. The length of life is probably about ten to fifteen years, but longer periods up to seventy years have been suggested. Many kinds of fish, including sticklebacks, serve as hosts for the glochidia.

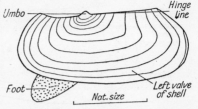

FIG. 182.—Young Swan Mussel (*Anodonta*).

Genus Pseudanodonta.—There is only one species, *P. minima*, which closely resembles *Anodonta*. It is only found in the Thames and Severn basins and some canals of Yorkshire and the Midlands.

Genus Unio.—This is easily distinguished from *Anodonta* and *Pseudanodonta* because on the outside of the shell there is a definite depression in front of the umbones (i.e. between the blunt end of the shell and the umbones). Also another

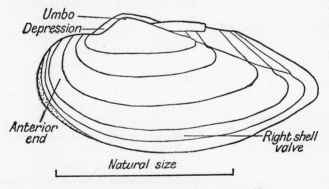

FIG. 183.—Shell of the Fresh-Water Mussel *Unio*.

difference is that on the interior of the valves along the hinge line are three pairs of " hinge-teeth " (see Fig. 184). Two of the three species (*U. pictorum* and *U. timidus*) are found in the same kind of places as *Anodonta*, but they only occur in

England and Wales. The third species is the Pearl Mussel (*U. margaritifer*, which has recently been placed in a separate

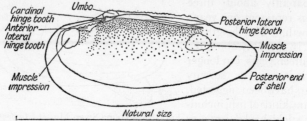

FIG. 184.—Right shell valve of the Mussel *Unio* viewed from inside.

genus, that of *Margaritifer*). The pearl mussel is found in certain rivers of north and west England, in Scotland, and

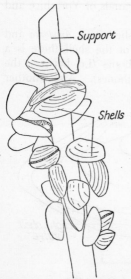

FIG. 185.—A group of Fresh-Water Mussels (*Dreissensia polymorpha*).

Ireland. It prefers quick-flowing rivers with clean water, a low lime content (soft water), and a gravel or sandy bed in which it can burrow. Its glochidia attack trout or minnows, but nothing is known of where or how the young live after leaving the fish until they are about 5 cm. long. The pearls which are obtained from this fresh-water mussel have considerable value. They are formed in the same way as those of the marine oysters,—concentric layers of pearly material are formed round some foreign body, such as a sand grain, which gets inside the animal's shell.

The last mussel, *Dreissensia polymorpha*, differs in some respects from all the others. In appearance the adults are like the sea mussels *Mytilus* (Fig. 185), and like them they attach themselves in groups to any suitable object, such as submerged palings, piers, or even the shells of other fresh-water mussels by means of chitinous threads with a sticky quality. They are found in docks, canals, and reservoirs, but not often in natural stretches of fresh water. The eggs hatch into a young larva (Fig. 186A) which has a crown of large cilia and which leads

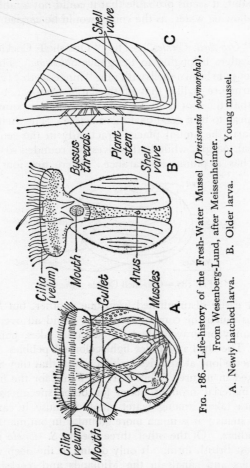

FIG. 186.—Life-history of the Fresh-Water Mussel (*Dreissensia polymorpha*). From Wesenberg-Lund, after Meissenheimer. A. Newly hatched larva. B. Older larva. C. Young mussel.

a free existence in the surrounding water like those of the marine mussels, instead of passing through this stage within its egg shell as is done by the others. This larva grows a shell (Fig. 186ʙ), and eventually attaches itself to some support by means of sticky threads, without ever going through a parasitic stage (Fig. 186ᴄ). As *Dreissensia* has this free-swimming larval habit, it seems probable that it could not establish itself in swift-flowing water, as the young would be carried down to the sea.

The " Orb-shell Cockles."—The " Orb-shell Cockles " are small bivalves belonging to the genus *Sphærium*. They have whitish or pale brown shells with nine " hinge-teeth " (Fig. 187). Their structure is like that of the large fresh-water mussels except that the foot is longer and rather more pointed, and they appear to be a good deal more active. The shell has the oldest part (the umbo) placed practically in the centre near the dorsal margin, while both valves are rounded so that the animal has a plump appearance. There are four species

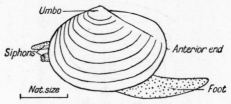

Fɪɢ. 187.—Orb-shell Cockle (*Sphærium*).

not very easily distinguished from one another, but *Sphærium corneum* is much the commonest and is found all over Britain in the sandy or gravelly beds of streams, lakes, ponds, and canals. The shell resembles light-coloured pebbles of about a centimetre long at first sight which means that they are often overlooked in nature. They may be found for the first time among material brought home for other specimens. It is sometimes taken among water-plants because it can climb up their stems ; it is much more globular in outline than any of the others. Of the other three species *S. rivicola* is much the largest British form. It only occurs in the deep water of slow rivers and canals in the Midlands and Yorkshire. *S. pallidum* (*ovale*) has a very pale coloured shell oblong in shape. It is only found in canals and does not possess active climbing habits. It is particularly abundant in the Manchester canals.

S. lacustræ is not very common but it is found in a variety of types of fresh water, including dew-ponds and ponds which often dry up. In the region of the umbo it has well-marked umbonal caps.

All the members of this genus are hermaphrodites. They do not lay their eggs in the water but keep them inside the shell where they hatch and develop into small editions of the parents before they are set free. About six or seven young are usually produced at a time.

" *Pea-shell Cockles*."—These all belong to the genus *Pisidium*. They are smaller than the " Orb-shell Cockles," have the same number of hinge-teeth, and in many species the umbo is not in the middle of the dorsal margin but rather behind it, giving the shell a lop-sided appearance. If a live specimen is examined in a dish of water contain-

FIG. 188.—The Pea-shell Cockle (*Pisidium*).

ing sand or gravel so that it can take up its natural position, it will be found to have only one siphon, while the genus *Sphærium* has two (Fig. 188). There are fifteen British species of " Pea-shell Cockles " which can only be identified by an expert, but several are very common in sand or mud at the bottom of almost any type of fresh water, including marshes, ditches, and mountain tarns. Like *Sphærium* they are hermaphrodite and viviparous, producing fully developed young.

Books for further information :—

British Snails, by A. E. Ellis. Pub. Oxford, 1926.

The Habitats of Fresh-water Mollusca in Britain. A paper by A. E. Boycott in the Journal of Animal Ecology (1936), Vol. 5, No. 1. (This gives a complete list of all the British fresh-water Mollusca, with maps showing their distribution, also much other valuable information.)

The Fresh-water Mussel. Chap. 6 in The Natural History of Some Common Animals, by O. H. Latter. Pub. Cambridge, 1904. (This gives a good account of the structure, habits, and life-history of the Swan Mussel (*Anodonta cygnea*).

For additional books, see p. 288.

CHAPTER 12

THE POLYZOA OR MOSS ANIMALCULES

THERE are many common marine Polyzoa, some of which
are known as " sea-mats " ; the fresh-water forms are less
prominent, but they are not really uncommon. The fresh-
water species are all colonial, and perhaps several hundred
of the individual " polyps " are found aggregated together.
The individuals look superficially like Hydras, but further
investigation shows that they are very different and much
more complex (Fig. 189A). The members of a colony are
connected together and protected by a strong sheath secreted
by the animals. The most prominent feature of the individual
is the *lophophore* (Fig. 189B), as the horse-shoe shaped crown
of tentacles is called. The lophophore is retractable, and if
the animal is disturbed it is withdrawn into the protective
covering and the polyzooan looks quite shapeless. In a few
minutes the animal will probably expand again. The tentacles comprising the lophophore are ciliated, and the cilia
make currents which collect the small organisms used as
food, and transport them to the mouth. Polyzoa possess
a proper gut, which terminates in an anus ; ovaries and
testes (they are hermaphrodites) ; and a rudimentary nervous
system. The fresh-water polyzoa reproduce asexually by
budding—a colony starts with one individual and " grows "
in this way. They also produce " statoblasts," which are
special buds covered with a resistant wall ; the statoblasts
live over the winter in a dormant condition, and germinate
in spring to start new colonies.

The three types most commonly found in fresh water are
described below :—

Genus Cristatella. Cristatella mucedo (the only member of

the genus) has colonies which may be 5 cm. long (Fig. 189c).
The whole colony is able to move about at a speed of perhaps
10 cm. a day. The form of the colony is quite characteristic
and unmistakable ; it is greyish-green in colour, and when
alive looks like a piece of furry jelly ! *Cristatella* occurs in
clear ponds and lakes, and is found on the upper sides of
plants and stones, for it " likes " sunlight. The statoblasts
are also characteristic, with a row of curiously hooked spines.

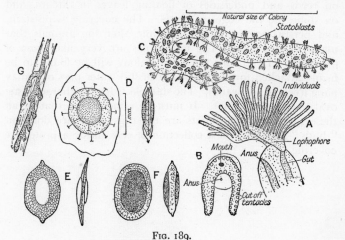

Fig. 189.
A. Individual *Cristatella*.
B. Lophophore diagram.
C. Colony of *Cristatella*.
D. Statoblast of *Cristatella*.
E. Statoblast of *Lophopus*. After Allman.
F. Statoblast of *Plumatella*. After Allman.
G. Sheath made by *Plumatella* colony.

The two halves of the statoblast case come apart when it
germinates. Statoblasts are commonly found in collections
of débris from ponds in winter—this is often the first way in
which a pond is known to contain *Cristatella* (Fig. 189D).

Genus Lophopus. *Lophopus crystallinus* has individuals much
like *Cristatella*, but the colonies (which measure less than
1 cm. across) only contain about a dozen individuals connected
together in a piece of jelly. The colonies are often found on

duckweed, and will not be recognised until they have been in an aquarium for some time and the lophophores with their fine tentacles have been extended. *Lophopus* produces stato-blasts which are little discs, almost oval in shape, thickened in the centre (Fig. 189E).

Genus Plumatella. *Plumatella repens* appears to be the commonest British fresh-water species. The protective covering showing its characteristic branching form (Fig. 189G) is found on reeds and undersides of floating leaves in ditches, and on stones at the edges of rivers. The covering is resistant, and persists some considerable time after the animals that made it have died. The individuals are very like those of *Cristatella*. In streams the whole colony will be found to be firmly attached to the stones, but in still water part of it may hang freely in the water. The statoblasts have a characteristic oval shape (Fig. 189F). It must be remembered that in all these species the statoblasts are not eggs—they are dormant " buds " consisting of a collection of cells inside a protective covering.

CHAPTER 13

THE WHEEL ANIMALCULES (*ROTIFERA*)

PRACTICALLY all the Wheel Animalcules or Rotifers are inhabitants of fresh water, and they are found in any type of water, from that of puddles and rain gutters to large lakes. All the species are microscopic, though the larger ones, which may be 2 mm. long, are visible to the naked eye as small whitish or pinkish bodies. In spite of their small size they have a complicated structure, and are made up of many different kinds of cells. They have always been very favourite objects for study to anyone interested in microscopy.

The main distinguishing feature of a rotifer is its complicated feeding and locomotor organ which is situated at the anterior end of the body (in a very few cases it is absent). This is called the *Corona*, and it consists of long vibrating cilia arranged in various ways among the different families, which by their movement cause a current of water bringing food to flow towards the mouth, at the same time as they propel the animal through the water. In a common group of rotifers (the *Bdelloida*) this apparatus is in the form of two flat discs which give the appearance of revolving wheels when the cilia are in motion. These were the first rotifers to be described, and they give the name of " Wheel Animalcules " to the group. The body of a Rotifer is usually cylindrical (see Fig. 190), but nearly spherical or very thin and attenuated forms also occur ; they are covered by an elastic skin, but this is sometimes hardened so that it maintains a definite shape, when it is called a *lorica*. The corona is expanded at the head end during locomotion and feeding, while at the opposite end the body is prolonged into a stalk or what is more often called the " foot," which generally ends in two pointed toes (see Fig. 190A, E). Both corona and foot can be retracted to some extent inside the main part of the body. In many forms the body appears to be divided into a number of segments, and there may be longitudinal markings as well.

Inside the skin there may be seen by transparency many of the internal structures while the animal is alive. Some of the less active kinds can be examined mounted in water in the ordinary way, on a slide under a cover-glass ; but for the small very active kinds it is better to slow down their movements by mounting them in a solution of gum in water, or if they are very common in your culture they may be entangled in thread algæ by placing some of these on the slide. There is a well-developed alimentary canal leading from the mouth, which is near the centre of the ciliated feeding apparatus, to the cloacal opening which is just above the foot. Near the front end of the gut is a special masticatory organ (mastax) consisting of one pair of large toothed jaws ; these may be observed opening and shutting. The characters of the jaws are used for identifying the different families. Behind the jaws the gut narrows to a fine tube before it suddenly expands into a broad stomach where digestion of the food takes place. From the hind end of the stomach a straight intestine leads to the cloaca. If the species under examination is a plant feeder and the gut contains green plant remains, then it is usually easier to make out the various parts of the gut. In addition to a well-developed alimentary system these minute creatures also have muscles, excretory, reproductive and nervous systems. They have no blood or special breathing organs, since the surface of their bodies has sufficient area in proportion to their volume for breathing to be done through their skin. Most kinds have one red eye (some have two) near the anterior end. In relation with this is the brain, appearing as a clear mass of nervous tissue, and from it several nerves are given off to the other tissues. On the corona long stiff cilia may be present, these are probably sensory in function.

The breeding habits of rotifers are very interesting but rather difficult to follow on account of the animal's small size. First of all male rotifers are very few and degenerate ; in a number of families they have never been found. They are much smaller than the female, have generally no alimentary canal, and are therefore unable to feed so that their life is exceedingly short. The females (and almost every rotifer you find will be a female) produce several kinds of eggs ; large ones with thin shells which develop into females, others like them but much smaller which develop into males, and a different kind of egg which has a hard thick shell and which develops, sometimes after a long period of rest, into a female. This " resting egg " is a fertilised egg while the other two kinds develop without fertilisation. Generally the eggs are laid by the female but some species carry their eggs about attached to their bodies (see Fig. 194C), while in a few cases the eggs develop inside the body of the female in which case the young are born well developed.

A few rotifers are able to encyst themselves inside a thick protective skin ; in this dormant condition they can stand being dried up and are probably dispersed by the wind. If they are placed in water they rapidly absorb water, swell up and resume their former rotifer shape. In dried-up rain gutters small pink cysts of a common rotifer, *Philodina roseola* (see Fig. 190E), are often found.

During locomotion the " foot " of the rotifer may be used as well as the corona. In those forms, which swim freely through the water, movement of the foot from side to side causes the animal to execute a spiral course, or to glide round in circles. It also serves as an organ of temporary attachment as glands producing a sticky substance open on it. A whole large group of rotifers (the *Bdelloida*) carry out a creeping movement with expansion and contraction of their bodies after the manner of a " looper " caterpillar, or a leech.

The following descriptions of rotifer families is not an exhaustive account of all the rotifers to be found in Britain, but it is hoped that most of the commoner genera have been included.

Family Philodinidæ

Rotifers with a corona shaped like two wheels when expanded. They make movements resembling those of a

creeping leech or a " looper " caterpillar. Their bodies are
divided into segments of which several at the front and hind

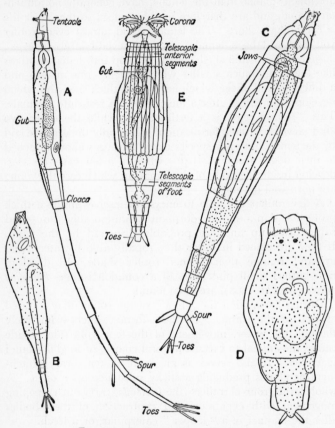

Fig. 190.—Rotifers. Family *Philodinidæ*.
A. *Rotifer neptunis*, extended, length 1 mm.
B. *Rotifer neptunis*, retracted.
C. *Rotifer*, creeping.
D. *Rotifer*, resting.
E. *Philodina*, length 0·3 mm.   Partly after Weber.

end can be withdrawn telescopically into one another. Males
unknown.

Genus Rotifer (Figs. 190A–D) has a pair of eyes very close to the front end of the body. The females are viviparous, producing young instead of laying eggs. One common species, *R. neptunis* (Fig. 190A), is long and slender : it is an amusing sight to watch it slowly extend itself as though it were a telescope being pulled out.

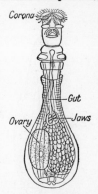

Genus Philodina.—Members of this genus are very similar to some of the previous genus, but the pair of eyes are much further back, just in front of the jaws. *P. roseola* is a pink species found in gutters and puddles (Fig. 190E). It encysts into a pink ball when its watery habitat dries up.

Genus Callidina.—This genus is eyeless (see Fig. 191) and is also capable of encysting. It is found chiefly among damp moss. There are a large number of different species.

Fig. 191.—*Callidina*, length 0·25 mm. Modified after Murray.

Family Notommatidæ

These rotifers have cylindrical soft bodies usually showing slight segmentation. The foot is not sharply marked off from the rest of the body and ends in two toes (occasionally only in one). The corona is generally narrower than the rest of the body, it is sometimes on the ventral side of the animal instead of being round the anterior end ; it consists of a ring of cilia round the edge, or is ciliated all over ; the mouth is not in the centre of the coronal disc but is nearer the ventral edge.

Genus Notommata has a large number of transverse folds across its body. There are many species, one of which (*N. werneckii*) lives as a parasite inside the filaments of various green algæ such as *Vaucheria* (Fig. 192).

Genus Diglena has a pair of eyes near the anterior end and its two toes are very long (Fig. 192C).

Genus Proales has a pair of eyes fairly far from the anterior end placed just in front of its jaws (Fig. 192B). One species,

Proales gigantea, is a parasite of the eggs of the Limnæa Water Snails (p. 240).

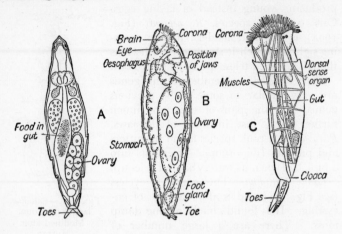

FIG. 192.—Rotifers. Family *Notammatidæ.*
A. *Notommata werneckii,* length 0·3 mm. From Wesenberg-Lund after Balbiani.
B. *Proales,* side view, length 0·3 mm. After Jennings and Lynch.
C. *Diglena,* length 0·2 mm. After Wesenberg-Lund.

Family Salpinidæ

In this family the skin is hardened to form a lorica (see p. 257) which gives these rotifers a definite shape ; the corona, and to a smaller extent the foot, can be retracted inside the lorica. Members of this family are free swimming and are common among plants in pools.

Genus Diaschiza (see Fig. 193A, B) contains at least ten British species. The lorica is here not very easy to see as it is less hardened than in some other genera.

Family Euchlanidæ

The lorica (see p. 263) is made of two plates in this family ; one which arches over the back of the animal is much curved, while the ventral one is flatter. They are joined at the sides sometimes by a furrow.

Genus Cathypna (Fig. 193c) is common in ponds.

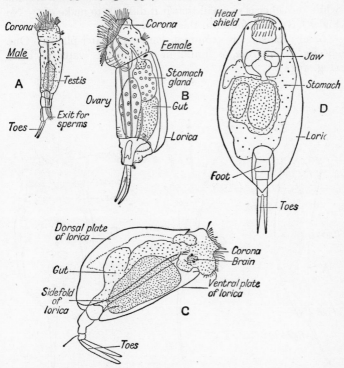

Fig. 193.

A and B. *Diaschiza*, male and female, male 0·1 mm. long. Modified after Dixon-Nuttall and Freman.

C. *Cathypna*, side view, length 0·15 mm.

D. *Colurus*, ventral view, length 0·1 mm.

Family Coluridæ

The lorica is made of one piece which sometimes only covers the dorsal side of the animal, usually with a rigid hood extending over the corona. Toes and foot fairly well developed.

Genus Colurus (Fig. 193D).

Family Rattulidæ

In this family the lorica is cylindrical and slightly curved ; of the two toes one is usually very long and shaped like a spine while the other is so small that it is difficult to make out. No males have been found in this family.

Genus Rattulus is common among water-plants. One species, *R. longiseta*, is found in the surface water of lakes (Fig. 194A).

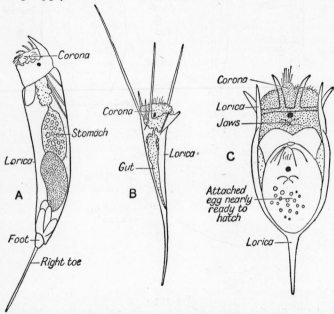

Fig. 194.—Three Rotifers from the surface water of Windermere.

    A. *Rattulus*, side view, length 0·5 mm.
    B. *Notholca*, side view, length 0·5 mm.
    C. *Anuræa*, ventral view, length 0·25 mm.

Family Anuræidæ

In this family the lorica is made up of one curved dorsal piece and a flat ventral one ; it possesses numerous spines, and as members of this family are found in large numbers in the surface water of lakes, the spines no doubt help to keep the animals floating ; in some, particularly the genus *Anuræa*, the shape of the lorica varies considerably at different times of the year.

Genera *Notholca* and *Anuræa* (see Fig. 194B, c) are British.

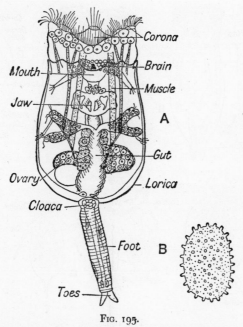

FIG. 195.

A. *Brachionus*, dorsal view, length 0·5 mm.   After Wesenberg-Lund.
B. Resting egg of Rotifer *Brachionus quadratus*, length 0·14 mm.   After Marks and Wesche.

Family Brachionidæ

In this family the animals have a distinct lorica made up of a curved dorsal and a flattened ventral plate. The eye-spots, if present, are fused into one. The foot is long and thick with rings on its surface, and ending in two small toes.

Genus *Brachionus* (Fig. 195A) is fairly common. The resting eggs have their surface covered with papillæ (Fig. 195B).

Family Hydatinidæ

Large cylindrical rotifers with the corona on the ventral surface bordered at the anterior edge by groups of stiff sensory cilia. Foot ending in two small toes.

The genus *Hydatina* is shown in Fig. 196A, B.

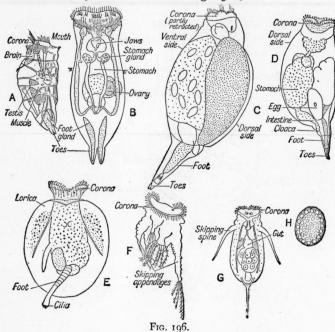

FIG. 196.

A. *Hydatina senta*, male, side view.   After Wesenberg-Lund.
B. *Hydatina senta*, female, ventral view, length 0·5 mm.   After Borradaile and Potts.
C. *Notops*, side view, length 0·7 mm.
D. *Harringia*, side view, length 0·25 mm.
E. *Pterodina*, ventral view, length 0·2 mm.
F. *Pedalion*, side view, length 0·7 mm.   After Hudson.
G. *Triarthra*, length 0·2 mm.
H. *Triarthra*, resting eggs.   After Rousselet.

Family Notopsidæ

Corona like that of the Hydatinidæ possessing stiff sensory

cilia. Skin slightly stiffened, body fat, foot clearly marked off from the body ending in two small toes (Fig. 196c).

Genus Notops is common. One species is found attached to the skin of surface swimming Crustacea, such as *Holopedium* and *Bythotrephes* (see p. 83).

Family Asplanchnidæ

The body of these rotifers is shaped like a sac, the foot is often absent, and the intestine may be lacking. In the genus *Harringia* (Fig. 196D) there is an intestine and a small foot.

Family Pterodinidæ

Corona has two wreaths of cilia round the edge. The body is enclosed by a lorica, and the foot, if present, does not end in toes.

The genus *Pterodina* (Fig. 196E) is exceedingly flattened ; the foot comes out from the ventral surface, it is ringed, and ends in a tuft of cilia.

Family Pedalionidæ

Corona with two rings of cilia round the edge. There is no lorica or foot but they possess curious appendages with which they skip. Representatives of the genera *Pedalion* and *Triarthra* are shown in Figs. 196F, G.

Family Melicertidæ

Nearly all the members of this family live fixed to some object by their foot, or inside a tube which they construct themselves.

The genus *Melicerta* has a corona made up of four lobes with large cilia round the edge. Its tube is constructed of small pellets (Fig. 197A).

Family Floscularidæ

The corona is generally very large and bears long cilia round its edge. These rotifers live in tubes secreted by the cells of the skin. The foot forms a disc which is attached to the far end of the tube. The genera *Floscularia* (which has its corona drawn out into lobes, Fig. 197C) and *Stephanoceros*

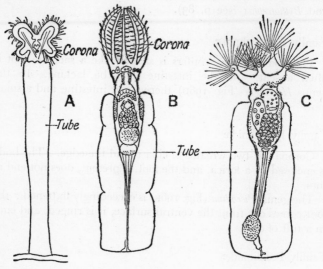

FIG. 197.

A. *Melicerta*, length 2 mm.
B. *Stephanoceros*, length 1 mm.   After Wesenberg-Lund.
C. *Floscularia*, length 0·7 mm.   After Wesenberg-Lund.

(whose corona is produced into pointed arms, Fig. 197B) occur in Britain.

To identify genera of Rotifera see

Fresh-water Biology, by H. B. Ward and G. C. Whipple. New York, 1918.

For general information—

Contributions to the Biology of the Rotifera, by C. Wesenberg-Lund.

Part I (1923), Part II (1930). Mem. Acad. Royale Sci., Copenhague.

APPENDIX TO THE ROTIFERA

The Gastrotricha or " *Hairy-backs.*"—These animals resemble the ciliated Protozoa in superficial appearance and habits, but their bodies are made up of many cells arranged in tissues to form muscles, a definite gut, a primitive nervous system, etc., and zoologists consider that they are allied to the rotifers. The largest known Gastrotrich is under half a millimetre long, so they are smaller than the larger Protozoa. Fig. 198 shows one of the commonest types, which occurs among algæ and other vegetation, particularly in upland streams. They are not uncommon, but being so small and insignificant they are seldom found. They scuttle about very like Ciliates, moving by means of cilia on their lower sides. They are all hermaphrodite, and the eggs, which are nearly a third the size of the parents, are laid among weeds. About a dozen British species are known.

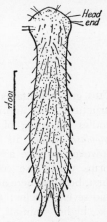

FIG. 198.—*Chætonotus*, a Gastrotrich.

CHAPTER 14

THE PROTOZOA

This group contains the animals with the most simple body structure. All animals which, when fully developed, have bodies consisting of one cell only (i.e. one *nucleus* and the *cytoplasm* under its control) are Protozoa. It is difficult, however, to give a simple definition which describes all the forms which are contained in the group for two reasons. First, because there are some members with complicated bodies and several nuclei ; these, however, are not truly multicellular like higher animals, and they are obviously closely related to other simpler Protozoa. (Some zoologists describe all Protozoa as " non-cellular " rather than " uni-cellular," but it is impossible to discuss the implications of this without having a detailed knowledge of the group.) Secondly, many lowly one-celled plants (Protophyta) are very similar to Protozoa, and only differ in the way they obtain their food. The Protozoa, being animals, feed by taking in complex food substances (i.e. by eating other animals, plants, or their decaying remains) which were ultimately produced by growing plants. The Protophyta feed for the most part like typical green plants—they contain chlorophyll which enables them to synthesise carbohydrates from carbon dioxide and water. There are " animals " which are intermediate between the Protozoa and the Protophyta like some *Euglenæ* which sometimes live like animals and at other times contain chlorophyll and live like plants.

Protozoa are small—almost all microscopic—and in their active stages they live always in contact with water, whether the water be free or part of the body of another organism. Such small creatures could not do otherwise, because they consist mainly of living protoplasm, the greater part of which is water. Such small creatures would dry up and die if exposed to the air, unless protected by resistant coverings (as they are in resting spores and cysts) which prevent any activity. Large numbers of Protozoa live in the sea, but many are found in fresh water. There are numerous parasitic species (for there is a supply of water inside the bodies of larger animals), and many of these are parasites of fresh-water animals.

As Protozoa are so small, as there are so many species, and as they are difficult to identify, it is impossible to deal with the group at all fully here. Since some species are continually being found by anyone using even a low-powered microscope, I propose to deal with the commoner and more distinctive kinds, so that it should be possible to get some idea of the relationships of most forms which are found. To go further you will require to be something of an expert on the group ; studying the books mentioned at the end of this chapter will help you on the first steps to become one !

To collect Protozoa it is only necessary to take home a little water from any pond rich in life. Large numbers of species will be found among water-plants, algal slime, and mud from the bottom. Often if some pond-water is left in a jar for several days—during which time the larger inhabitants may die and the whole may become unpleasant and smelly—then the Protozoan fauna will be found to have grown and multiplied. In fact this is the best method of obtaining many species (especially certain Ciliates), which are rare and difficult to find in nature.

The Protozoa are divided into four classes, three of which are distinguished by their methods of locomotion, and the fourth contains only parasites (there are some parasitic forms among all the other groups also). Members of all four classes are found in fresh water.

Against each Figure in this chapter there is a line to indicate the size of the animal. The line's length (when it is considered to be on the same scale as the animal) is given in " μ " (pronounced " mew "), which is the abbreviation for " micron " ; one micron is a thousandth of a millimetre, and is the unit of length normally used by microscopists. Thus $500\mu = \frac{1}{2}$

a millimetre, and an object 100μ long is just visible to the naked eye.

Class Flagellata

The first great sub-division of the Protozoa comprises the Flagellates. These animals are now considered to be the most primitive members of the phylum, while those contained in the next sub-division, the Rhizopoda (which includes the well-known *Amœba*), though superficially more simple in structure may yet have been derived from flagellate-like ancestors.

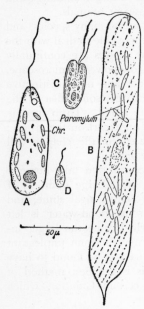

FIG. 199.

A. *Euglena viridis.*
B. *Euglena oxyuris.*
C. *Chilomonas.*
D. *Copromonas subtilis.*
All drawn to same scale.
Chr. = *chromatophore.*

The characteristics of a typical flagellate are shown in Fig. 199A, which depicts the form *Euglena viridis.* At the anterior end there is the whip-like flagellum — the feature which characterises the group. It is the locomotive organ, and by vibrating it in a definite but complex manner the *Euglena* pulls itself forward (with the flagellum in front) through the water. The flagellum is made from the protoplasm of the body of the animal and is in fact composed of specialised protoplasm. Some members of this group will temporarily lose their flagella and then grow new ones. The body of *Euglena* has on the outside a tough, elastic, and transparent "skin" which can change its shape a good deal, but in rest the body assumes an elongated pear shape. At the anterior end of the body there is an opening leading to a gullet. In allied forms food is taken in by the gullet, but it is doubtful whether *Euglena viridis* ever takes in solid particles of food. The gullet is used in this animal for excretion—a contractile vacuole near at hand collects fluid excretory matter from the body, and passes it out in this way. A small orange pigmented " eye-spot " is thought to be the point where sensitivity to light is concentrated. The most obvious features of the living animal are, usually, the numerous green *chloroplasts*. These enable the animal to synthesise carbohydrates like a plant. If, however, *Euglena viridis* is kept in water rich in food substances (i.e. rather putrid water), it absorbs these by their diffusion through its " skin." Under these circumstances the animal does not photosynthesise (for it obtains enough food directly), and the chloroplasts become colourless. The body also contains granules of food reserves, and a single nucleus. This nucleus is faintly visible in life, but very plain in specimens which have been killed and stained specially for microscopic examination.

Euglena viridis is commonly found in pond-water. There are many other common related forms, some of which are green and some colourless. A common colourless species is *Peranema ;* it resembles a *Euglena* without chloroplasts, and usually occurs in infusions made by soaking hay in water. The different species are very difficult to identify, but most of the larger flagellates generally seen belong to the order *Euglenoidina*. One interesting (green) species is *Euglena oxyurus* (Fig. 199B), which is very large, and is often found without any flagellum creeping about it in a peculiar manner. Under these conditions it is at first difficult to spot that it is even a protozoon—even experienced microscopists have been known to call it a worm ! In this species the most obvious feature is the presence of rod-like granules of paramylum, a carbohydrate food reserve.

Perhaps the commonest protozoon of all is *Chilomonas*. These forms are tiny flagellates, with a pair of flagella and a tougher " skin " than *Euglena* (Fig. 199C). They are colourless, and almost invariably appear in large numbers in foul water. When laboratory cultures of other species of Protozoa are kept for a few days it often happens that *Chilomonas* will entirely replace the original animal. Another common

18

form is *Copromonas subtilis* (Fig. 199D), which is almost always present in great numbers in frogs' or toads' rectums. It also occurs in foul water. This species has only one flagella, and is extremely small—only a third the size of *Chilomonas*. In the frog *Copromonas* seems to be a parasite which does no harm to its host.

Most Flagellata reproduce by dividing the body down the middle into two new individuals. Sometimes the flagella divide too, but in many forms one of the two daughter animals receives the original locomotor organs, and the other grows new ones. At times most of these animals conjugate—two adult animals come together and fuse (like two gametes of higher animals). The animal which results from the union does not, in most of these forms, divide again to form a large number of spores. It may divide to form only two individuals like a normal specimen.

There are large numbers of flagellated green unicellular plants—forms which invariably live by photosynthesis.* These are obviously closely allied to the animal flagellates. Most of these plants are solitary, like the common two-flagellated *Chlamydomonas*. However, some green forms produce colonies, small irregular masses of about sixteen individuals which swim about as one in the case of *Pandorina ;* a similar number spaced out in a flat jelly-like plate in *Eudorina ;* and finally *Volvox*, where hundreds of flagellate individuals are found to cover the outside of a gelatinous sphere which looks bright green due to their chlorophyll, and, being quite half a millimetre in diameter, is easily visible to the naked eye. These various colonial plant flagellates have two flagellæ per individual.

Most animal flagellates are very small, and few details of their anatomy are visible except under high magnifications and by experts. Of the smaller forms, the majority seen will be of the *Chilomonas* type, or of some not distantly allied species ; or else they will resemble *Copromonas*. The majority of larger flagellates will be Euglenoids. The remaining minority comprises hundreds of species, which, as has been said above, only experts can identify satisfactorily.

* The Euglenoids are often considered to be plants also.

Class Rhizopoda

During most of their active life all Rhizopods are "amœboid," and move about by means of *pseudopodia*, which are more or less irregular protrusions of the body cytoplasm. The description of the form *Amœba* will make this clearer (see Fig. 200A below). This section of the Protozoa was formerly thought to be the most primitive, because it contains forms which are apparently almost structureless bits of protoplasm. But it has been shown that certain of these, at some stage in their life, develop flagella like those found in the Flagellata. This happens most commonly in the immature stages, and it is now usually believed that the amœboid forms are degenerate descendants of flagellate ancestors.

The fresh-water Rhizopoda are divided into three easily defined orders. They all are definitely " animals," and do not show plant-like characters as do the flagellates. They mostly feed by ingesting solid particles of food, either smaller unicellular plants and animals, or their decaying remains. Rhizopods are not uncommon, but they are not anything like as frequently seen in random collections of pond-water as flagellates, or as members of that other class, the Ciliates. In fact, it may be some time before the beginner finds any specimens. *Amœba*, which is so well known as an examination " type," is particularly difficult to find in any numbers.

Order Amœbina

Rhizopods which are defined as having no shell or skeleton, and usually with rather blunt pseudopodia. *Amœba proteus* (Fig. 200A) may be taken as typical. It occurs in ponds, on the surface of the mud at the bottom, and sometimes among weeds. As mentioned above, it is never very common, and its body-form makes it difficult to see, particularly as it is nearly transparent. The following points are easily made out. The whole irregularly shaped body consists of two layers of living protoplasm, a thin outer clear ectoplasm, and an inner granular endoplasm. After being disturbed, say by being picked up in a pipette, *Amœba* may contract into a lump and not move for several minutes: contracted it is very difficult to find. When this animal moves, the inner endoplasm first flows out into a lobe—a pseudopodium or " false foot "—and the rest of the amorphous body follows the lobe. The nucleus is visible, but not very obviously, in the live animal. A clear spherical contractile vacuole pulsates regularly. As it expands it fills with superfluous water, perhaps containing some dissolved excretory substances.

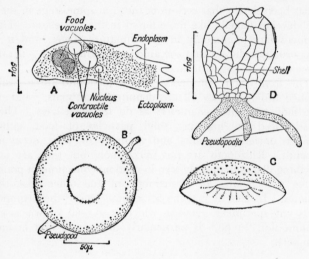

FIG. 200.

A. *Amœba.* B. *Arcella* from above.
C. Shell of same. D. *Difflugia.*

When it contracts the liquid is expelled by a temporary break in the ectoplasm. Many of the larger granules are bacteria or debris which have been taken in as food and are in process of being digested. The surrounding protoplasm produces digestive enzymes, the food-substances are absorbed, and the residue is expelled from the body. *Amœba* usually reproduces simply by dividing into two. The nucleus divides first, and the body then splits and the two parts move away. Under unfavourable conditions, *Amœba* may encyst and secrete a strong protective covering. The body inside then divides into many small parts, each containing a nucleus, and each part can, under favourable conditions become a new animal.

Amœba proteus is the largest species and that most commonly found. It can be cultured in shallow dishes of water to which two or three well-boiled wheat grains have been added. Many other *Amœbæ* occur in fresh water, some of which have a flagellate stage in their life-history.

Order Foraminifera

These Rhizopods have a shell. Particularly in the sea there are very complex forms, but the two commonest fresh-water types are just like amœbæ which shelter inside a definite type of protective covering.

Arcella (Fig. 200B, c) occurs in ponds, among the mud particularly if this is rich in decaying vegetable matter. It also occurs in marshy places. The smooth shell is made of chitin, and is secreted by the animal. *Difflugia* (Fig. 200D) occurs in similar habitats. This animal secretes a thin shell which is then covered over with sand and débris. Multiplication is by division into two, but before it divides the parent secretes a new shell at the mouth of its own ready to receive the daughter (half) ; the two then separate.

Arcella and *Difflugia* are the two commonest fresh-water Foraminifera. Many other species also occur, bearing cases of various shapes and materials.

Order Heliozoa

These forms are characterised by having pseudopodia which are slender and stiffened by an axial filament—such pseudopodia are called *axopodia*. When Heliozoa move they roll rather than creep. Some species have a siliceous skeleton.

Actinophrys sol (Fig. 201A) is the commonest Heliozoan. It is like a tiny prickly ball, and occurs particularly among decaying leaves. A nucleus, food vacuoles, contractile vacuoles, and pseudopodia are all visible. A fairly common species, *Actinosphærium*, looks very like *Actinophrys* except that it contains many nuclei instead of one and it is also twelve or more times as big. Specimens have been found up to a millimetre in diameter. *Clathrulina*, which occurs in ponds, is the commonest Heliozoan with a siliceous skeleton. The skeleton (Fig. 201B) looks like a stalked golf-ball, and the animal sits inside and sticks its pseudopodia out through the holes. Division into two individuals takes place inside the lattice-like skeleton, and one half of the protoplasm then flows out through an opening and secretes a new stalk and skeleton, perhaps still attached to the parent.

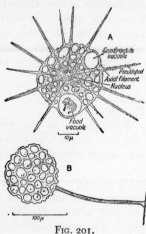

FIG. 201.

A. Heliozoa *Actinophrys sol.*
After Bronn.
B. *Clathrulina*.

Class Ciliophora

The members of this class are characterised by the possession of cilia as their means of locomotion. Cilia are like smaller and very numerous flagella (see p. 272 above). The ciliates are often larger than most other Protozoa, they are common, they multiply readily in rather foul water in the laboratory, and so they are very frequently seen by anyone using a microscope. Some individuals reach a millimetre in length, and others form colonies which are readily visible to the naked eye. All ciliates are animals, taking in food usually through a definite gullet leading in through the ectoplasm, though some forms absorb their nourishment through the general surface of the ectoplasm. When ciliates appear green in colour, this is due to their containing numerous tiny unicellular plants living in their bodies in a state of symbiosis. That is to say, the ciliates give the tiny plants protection and necessary salts, while the plants supply the ciliates with some of the products of their photosyntheses. If you wish to obtain ciliates to study, first make a nutrient infusion by adding boiling water to a little hay. Let this stand for several days, and then add pond-water which contains a few of these animals. They will probably multiply rapidly in the infusion. Actually, it is often possible to get rich cultures by just leaving some pond-water containing a little weed to " go bad." Under all these conditions the protozoans become more numerous and easier to find than in the wild state. Cultures should be examined often, as generally you get a rotation of species, and one species after another seems successively to become abundant and then die out as it is replaced by the next. The descriptions given below are for the species of ciliates most commonly found. There are of course many others not mentioned here.

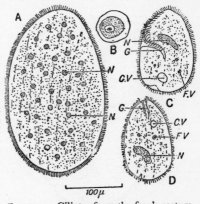

FIG. 202.—Ciliates from the frog's rectum.

A. *Opalina ranarum.*
B. *Opalina ranarum* cyst.
C. *Nyctotherus cordiformis.*
D. *Balantidium entozoon.*
C.V. = Contractile vacuole.
F.V. = Food vacuole.
G. = Gullet.
N. = Nucleus.

Order Holotricha

Ciliates with cilia more or less uniform in size and in distribution over the body.

Sub-order Prociliata. No mouth present.

Opalina ranarum (Fig. 202A).—If the rectum of a frog is cut open, numerous ciliates will nearly always be found among the contents. The commonest and largest is *Opalina ranarum*, an animal which may reach a millimetre in length. They are visible to the naked eye, and may be picked out with a needle. *Opalina* is thin and flat, covered all over with cilia, with a tough and elastic ectoplasm (such as is found in all ciliates) which allows a good deal of change in shape. There are numerous (perhaps fifty) small nuclei. *Opalina* forms cysts (Fig. 202B) which are excreted into the water, swallowed by another frog, and so cause further infestation. *Opalina* and the other protozoan parasites in the frog's rectum seem to do the frog no harm.

Sub-order Gymnostomata

These forms have a definite mouth, but they do not have gullets lined with large cilia like many of the forms described later, so the gullets are not seen easily.

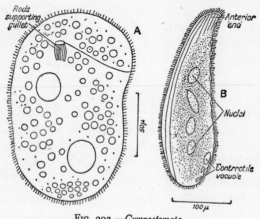

FIG. 203.—*Gymnostomata.*
A. *Chilodon.*   B. *Loxophyllum.*

Chilodon (Fig. 203A) has numerous contractile vacuoles, and occurs mainly in stagnant water among algæ. Its gullet is supported by curious little rods. *Loxophyllum* (Fig. 203B) is characterised by a crenulated border ; it occurs in ponds.

Sub-order Vestibulata

The mouth of these forms is kept permanently open, and the gullet is lined with cilia and consequently easily seen. The familiar *Paramœcium* (Fig. 204) belongs here. It is a large slipper-shaped organism commonly found and easily bred in infusions. It possesses all the typical ciliate features described above, and detailed descriptions of its anatomy, are given in most elementary zoology text-books. *Colpidium* (Fig. 205) is much smaller than *Paramecium*, and even more commonly found. Its body possesses obvious longitudinal striations.

Order Heterotricha

These forms all have a wreath of extra long cilia round the mouth, while the rest of the body is usually uniformly ciliated.

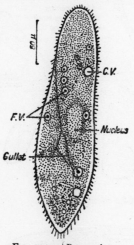

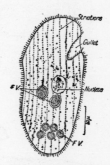

FIG. 204.— *Paramecium*.

C.V. = Contractile vacuole.
F.V. = Food vacuole.

FIG. 205.—*Colpidium*.

There are two other ciliates commonly found, together with *Opalina* (see p. 279 above), as parasites in the rectum of the frog. They are *Nyctotherus* (Fig. 202C) and *Balantidium* (Fig. 202D). These are both heterotriches. *Nyctotherus* has its gullet at one side, and *Balantidium* its gullet at the anterior end. These two ciliates are considerably smaller than *Opalina*, and they are seldom as common, and sometimes one and not the other is present. *Spirostomium* (Fig. 206B) occurs in ponds. It is long and very slender, and may reach a length of over 2 mm. *Stentor* is another large ciliate. It is much broader at the oral end than at the other extremity, by which the animal commonly attaches itself to plants or to the sides of an aquarium. Colonies occur (Fig. 206A), in which the lower parts of all the animals are covered with a mucilaginous

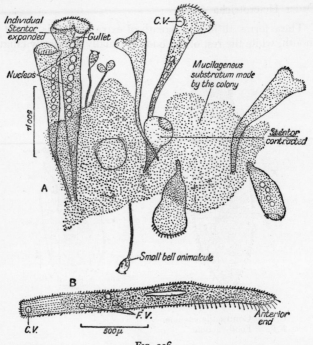

Fig. 206.

A. Colony of *Stentor polymorpha*, including individuals of the Bell Animalcule (*Vorticella*).
B. *Spirostomum*.

C.V. = Contractile vacuole.
F.V. = Food vacuole.

substance which they secrete. Individuals often swim away from the colony; as they do so they take on a barrel-like shape. The commonest *Stentor*, *S. polymorpha*, is green due to symbiotic organisms inside its body. Another form, *S. cœruleus*, which is blue, is often found. Although *Stentor* is only a protozoan, it is large enough to be able to eat small crustaceans such as young water-fleas. A common but very much smaller Heterotrich is *Halteria* (Fig. 207). Very little of this animal's anatomy is visible during life, because it spins and leaps—the movements are quite characteristic. *Halteria* shows the typical circle of cilia (which cause the spinning) and it also has long bristles—by a sudden movement of which the whole animal " leaps."

Order Hypotricha

The members of this order are characterised by having their cilia fused together into stiff bristles or cirri. These

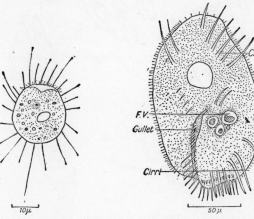

FIG. 207.—*Halteria*.   FIG. 208.—*Stylonichia*.

cirri occur mainly on the underside, as in *Stylonichia* (Fig. 208), and the animals run about on the bottom of the ponds using the cirri as " legs." *Stylonichia* occurs commonly, and is easily recognised. A rather similar form called *Keroma* is sometimes found as an ectoparasite scuttling over the body of *Hydra* (see also p. 285 below).

Order Peritricha

In these forms, most of which are normally permanently fixed by the end opposite to the mouth (aboral end), the

cilia are reduced to a long ring round the mouth, the rest
of the body being naked. *Vorticella* (Fig. 209A, B, C), the Bell

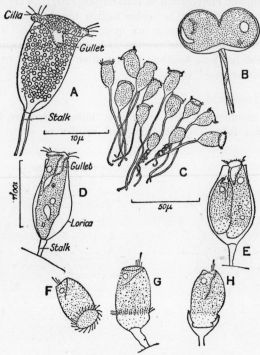

Fig. 209.—Peritricha.

A, B, C. The Bell Animalcule (*Vorticella*).
A. Mature individual.
B. Individual dividing.
C. Part of colony.
D, E, F, G, H. *Cothurnia*.
D. Mature individual.
E. Individual dividing.
F. Free swimming stage.
G, H. Metamorphosis of free swimming stage.

Animalcule, has a long stalk which can contract suddenly
in a spiral manner. *Vorticella* is often found in colonies,
probably caused by the divisions of one individual. There

is a free-swimming stage (similar to Fig. 209F) which disperses the species. Some *Vorticellæ* (not all) are green, and a colony is visible to the naked eye as a green disc perhaps several millimetres across. They often develop in a few days on the walls of aquaria. When disturbed the disc contracts to about a third its original size quite suddenly, and then a few seconds later it slowly expands again. This is due to the individuals all contracting and expanding their stalks. *Vorticella* is really social rather than colonial, for there is no organic connection between mature individuals. In the allied *Epistylis*, however, real branching colonies of perhaps several hundred vorticella-like (colourless) individuals are found. *Cothurnia* (Fig. 209D) possesses a cup-like outer sheath called a lorica, inside which it sits. There is a free-swimming stage (Fig. 209F) which attaches itself by the aboral end (most of the cilia which are visible are in this stage aboral). Then it forms a stalk, loses the aboral cilia, and secretes the lorica (Fig. 209G, H). The whole process may take an hour. Multiplication can take place by one

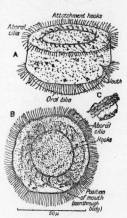

FIG. 210.—*Trichodina*.

A. From side.
B. From below, as usually seen.
C. On tentacle of *Hydra*, smaller scale.

individual inside its lorica dividing into two (Fig. 209E), when one half then swims off to settle down elsewhere. *Cothurnia* is often found living on the body of some crustacean like *Diaptomus*. It is then an epiphyte—it does no harm to the crustacean. On *Hydra* (see p. 16) a parasitic "peritrich" called *Trichodina* (Fig. 210) commonly occurs. It either scurries about the *Hydra*, or sits anchored by the prominent hooks on the aboral end.

Class Sporozoa

Protozoa which spend the main part of their lives as internal parasites. Some sporozoans occur in many fresh-water animals, but they are usually difficult to see, and they are seldom found by the ordinary collecting methods. One exception is *Glugea anomala*, which attacks the stickleback.

FIG. 211.—The Sporozoon *Glugea anomala* on stickleback.

The result is one or more white nodules (Fig. 211) on the side of the fish. Inside of these nodules the parasite gives rise to thousands of tiny spores which are eventually set free to infect other sticklebacks. It is usual to find many of the sticklebacks in a pond or canal infected, or else none at all. Most of the other Sporozoa have very complicated life-histories, which it is impossible to discuss in a book of this nature. They are not, however, commonly found at all by the amateur.

Books for further reference :—

General.

> Wenyon, C. M. (1926). Protozoology. Vols. 1 and 2. London.
>> Primarily medical, but is an excellent introduction to the group ; Sporozoa very fully described.
>
> Ward, H. B., and Whipple, G. C. (1918). Freshwater Biology. New York.
>> Devotes 90 pages, with many illustrations, to American forms.

Flagellata.

> Die Süsswasserflora Deutschlands. Hefte 1, 2. Jena, 1913.
>> Describes Protozoa and Protophyta of Germany (in German).
>
> Kent, W. Saville (1880). A Manual of the Infusoria. Vols 1-3. London.
>> Out of date, but still very useful. Vol. 3 entirely plates.

Rhizopoda.

Cash, J., and Wailes, C. H. (1904-1921).   The British
Freshwater   Rhizopoda   and   Heliozoa.   Vols.   1-5.
Ray Society, London.
With the aid of this book and its excellent illustrations
even a beginner can identify many species.

Ciliophora.

Kent, W. Saville (see above).
Roux, J. (1901).   Faune infusorienne des eaux stagnantes
des environs de Geneva.   Geneva
Incomplete, but useful (in French).

Sporozoa.

Wenyon, C. M. (see above).

The books mentioned above give extensive lists of references
to more detailed works.

Further books on Dragon-flies :—

The Generic names of the British Odonata with a check list of the species. J. Cowley (1935), *Generic Names of British Insects.* **3,** 45. Royal Entomological Society, London (1949).

The Dragon-flies of the British Isles. C. Longfield, (1949) London. Enlarged Edition.

Handbooks for the Identification of British Insects. Odonata F. C. Frazer, (1949). Royal Entomological Society. London.

Further books on Nymphs :—

Handbook for the Identification of British Insects. *Ephemeroptera* by D. E. Kimmins (1950). Royal Entomological Society, London.

Descriptions of the Nymphs of the British Species of Cloëon, Procloëon, and Centroptilum (*Ephemeroptera, Baëtidæ*) by T. T. Macan (1949). The Entomologist's Monthly Magazine, **85,** 222.

Descriptions of some Nymphs of the British Species of the Genus Baëtis by T. T. Macan (1950). Trans. Society British Entomologists, **10,** 143.

The taxonomy of the British Species of Siphlonuridæ by T. T. Macan (1951). Hydrobiologia, **3,** 84.

Further books on Mollusca :—

A Key to British Fresh- and Brackish-Water Gastrapods, by T. T. Macan and R. Douglas Cooper (1949). Freshwater Biological Association. Scientific Publication No. 13.

The Identification of the British Species of Pisidium, by A. E. Ellis (1940). Proc. Malacalogical Society, **24,** 44.

Freshwater Bivalves (Mollusca), A. E. Ellis (1946). Linnean Society London Synopsis of the British Fauna No. 4.

INDEX

A

Acanthocephala, 45.
Acentropus, 181.
acetabula, 220, *224.*
Acilius, 159.
Actinophrys sol, 277, *278.*
Actinosphærium, 277.
Aëdes, 204, 206.
Æolosoma, 50.
Æolosomatidæ, 50.
Æschnidæ, *124,* 126.
Agabus, 160.
Agapetus, *192,* 193.
Agraylea, *191.*
Agridæ, 126.
Agrion, 125.
Agriotypus, *182.*
Alder-flies, 150 et seq.
Amnicola, 236.
Amœba proteus, 275, *276.*
Amphipeplea, 241.   See Myxas.
Amphipoda, 107.
Anabolia, 190.
Anacæna, 176.
Ancylastrum, 237.
Ancylus, 237.
Anisoptera, *124,* 125, 126.
Anodonta, *244, 245, 247, 249.*
Anopheles, 204, 205, 206.
Anthomyiidæ, 214, 215.
Anuræa, 264.
Anuræidæ, 264, 265.
Aphelochirus, 143.
Aplecta, 238.
Apparatus for collecting, 7.
Apus, 72.
Arachnids, 216 et seq.
Arcella, 276, 277.
Argulus, 103, *104.*

Argyroneta, 216, 217.
Arrhenurus, 226.
Artemia, 72.
Asellus, *105, 106.*
Asplanchnidæ, 267.
Astacus, 110, *111, 112, 113.*
Athericera, 213.
Autotomy, 114.

B

Bætis, 133, 136, *138.*
Balantidium entozoon, 279, 281.
Bdellocephala punctata, 24.
Bear Animalcules, 227.
Bears, Water-, 227.
Beetles, 153 et seq.
Bell Animalcule, *284.*
Berosus, *175,* 176.
Bithynia, *233.*
Bivalves, 244.
Black-flies, 206 et seq.
Blood-worms (*Chironomous*), 210, 211.
Bosmina, 80.
Bosminidæ, 80.
Brachionidæ, 265.
Brachionus, 265.
Brachycera, 212.
Braconidæ, 183.
Brychius, 163.
byssus, *247,* 248.
Bythotrephes, 82, *83.*
Bythinia, 144.

C

Caddis-flies, 184 et seq.
Cænis, 137, *139.*